CHEMISTRY RESEARCH AND APPLICATIONS

PHENYLALANINE

DIETARY SOURCES, FUNCTIONS AND HEALTH EFFECTS

CHEMISTRY RESEARCH AND APPLICATIONS

Additional books in this series can be found on Nova's website under the Series tab.

Additional e-books in this series can be found on Nova's website under the e-book tab.

CHEMISTRY RESEARCH AND APPLICATIONS

PHENYLALANINE

DIETARY SOURCES, FUNCTIONS AND HEALTH EFFECTS

DARLENE WILLIAMSON
EDITOR

nova publishers
New York

NOTICE TO THE READER

The Publisher has taken reasonable care in the preparation of this book, but makes no expressed or implied warranty of any kind and assumes no responsibility for any errors or omissions. No liability is assumed for incidental or consequential damages in connection with or arising out of information contained in this book. The Publisher shall not be liable for any special, consequential, or exemplary damages resulting, in whole or in part, from the readers' use of, or reliance upon, this material. Any parts of this book based on government reports are so indicated and copyright is claimed for those parts to the extent applicable to compilations of such works.

Independent verification should be sought for any data, advice or recommendations contained in this book. In addition, no responsibility is assumed by the publisher for any injury and/or damage to persons or property arising from any methods, products, instructions, ideas or otherwise contained in this publication.

This publication is designed to provide accurate and authoritative information with regard to the subject matter covered herein. It is sold with the clear understanding that the Publisher is not engaged in rendering legal or any other professional services. If legal or any other expert assistance is required, the services of a competent person should be sought. FROM A DECLARATION OF PARTICIPANTS JOINTLY ADOPTED BY A COMMITTEE OF THE AMERICAN BAR ASSOCIATION AND A COMMITTEE OF PUBLISHERS.

Additional color graphics may be available in the e-book version of this book.

Library of Congress Cataloging-in-Publication Data

ISBN: 978-1-63482-501-6

Published by Nova Science Publishers, Inc. † New York

CONTENTS

Preface		**vii**
Chapter 1	Biological Roles of Phenylalanine and Its Derivatives; The Genetic Disorder Phenylketonuria and Its Possible Treatment; Potential Separation Methods of Phenylalanine Enantiomers and Biological Functions of D-Phenylalanine *Marek Graff*	**1**
Chapter 2	Phenylketonuria's Impact on Physical Growth in a Spanish Cohort *Luis Aldámiz-Echevarría, Marta Llarena, Fernando Andrade, María L. Couce, Domingo González-Lamuño, María A. Bueno, María C. García, Jaime Dalmau, Isidro Vitoria, Javier Blasco, Carlos Alcalde, David Gil, Mónica Ruiz, Luis Peña-Quintana, María A. Ruiz, David González and Felix Sánchez-Valverde*	**27**
Chapter 3	Role of Phenylalanine in Viticulture and Enology *P. Santamaría, R. López, J. Portu, L. González-Arenzana, I. López-Alfaro, A. R. Gutiérrez and T. Garde-Cerdán*	**49**
Chapter 4	Penetration of L-Phenylalanine and Amino Acids into Hair Keratin Fibers *Akio Kuzuhara*	**71**
Index		**93**

PREFACE

Phenylalanine is an α-amino acid with the aromatic ring in the molecule, similar to other amino acids containing aromatic rings, like tyrosine and tryptophan. Phenylalanine is classified as nonpolar because of the hydrophobic character of the benzyl group as the side chain. This book discusses the dietary sources of phenylalanine. It also discusses it's functions and effects on health.

Chapter 1 - Phenylalanine (Phe) is an α-amino acid with the aromatic ring in molecule, similar to other amino acids containing aromatic rings, like tyrosine and tryptophan. Phenylalanine is classified as nonpolar because of the hydrophobic character of the benzyl group as the side chain. This amino acid is used as a precursor in a tyrosine synthesis and other compounds containing a six-membered aromatic ring, like dopamine, and compounds known as catecholamines: epinephrine (adrenaline), norepinephrine (noradrenaline), and melamine (the skin pigment). Phenylalanine was discovered in yellow lupine by Schulze and Barbieri in 1879, and synthesized by Erlenmeyer and Lipp three years later. In nature, phenylalanine is found in the breast milk of mammals. Excess of phenylalanine in the body is known as phenylketonuria (PKU) – it is a genetic disorder, related to lack of phenylalanine hydroxylase (PAH), an enzyme catalyzing the hydroxylation of phenylalanine to tyrosine (in aromatic ring). PKU is described as an autosomal recessive inborn error of metabolism. Test of the Phe level in blood is carried out for new born babies, for eliminating potential PKU; lack of control and possible of PKU can cause of brain damages. Patients with PKU have to control their intake of phenylalanine. This amino acid is also used in nutritional applications like aspartame – the artificial sweetener. From another hand, D-phenylalanine is found in proteins in small amounts, especially in processed ones, but biological roles of D-amino acids remain unclear. However, it is known the pharmacological activity of D-phenylalanine at the niacin receptor. Additionally, D-phenylalanine has also an antidepressant activity. It is known the fact that a cross the blood-brain barrier for D enantiomer is less efficient in comparison to L one. L-phenylalanine is synthesized on an industrial scale the bacterium *Escherichia coli* and used for medical, feed, and nutrition applications.

Chapter 2 - Phenylketonuria is an inborn error of metabolism affecting the phenylalanine metabolic pathway, which converts phenylalanine to tyrosine via phenylalanine 4-hydroxylase and its cofactor (6R)-L-erythro-5,6,7,8-tetrahydrobiopterin (BH_4). Phenylketonuria treatment is based on strict vegetarian diets, with very low phenylalanine intake and supplemented with phenylalanine-free formulas. This phenylalanine-restricted diet has proven to be effective in preventing the development of long-term neurological damage caused by phenylalanine accumulation. However, such diets have occasionally been reported to hinder normal development, since some individuals presented growth retardation. Furthermore, in an attempt to improve outcomes in these patients, several alternative approaches for the treatment of phenylketonuria have emerged, such as therapy based on administration of the cofactor BH_4. This treatment allows patients to consume a near normal diet, or at least, a less strict low-protein diet, which results in higher natural protein intake. However, little is known about how BH_4 treatment affects physical development.

Firstly, in order to evaluate the impact of the phenylketonuria diet on anthropometric characteristics (weight, height, body mass index and growth rate), the authors conducted a retrospective longitudinal study. The aim was to further explore the hypothesis of higher natural protein intake being associated with attaining improved physical outcomes. Anthropometric characteristics and nutrition were evaluated from birth to adulthood in a cohort of phenylketonuria and mild-hyperphenylalaninaemia patients, who were exclusively on protein-restricted diets, and were compared to the Spanish reference population. Patients with phenylketonuria showed growth impairment in early stages, with higher phenylalanine intakes being associated with improved developmental outcomes over this period. The author's results suggest that prescribing very stringent diets in early stages might predispose these patients to later growth retardation, with growth outcomes in adulthood being well below the 50th percentile for healthy subjects.

Secondly, the authors conducted a retrospective longitudinal study of anthropometric characteristics in a cohort of patients with phenylketonuria on BH_4 treatment and compared their developmental outcomes with those of a group of patients on a phenylalanine-restricted diet, in order to determine whether BH_4 treatment was associated with an improvement in growth development. Data were collected every 6 months over two periods of time (two and five years). No improvement was observed in the anthropometric variables in the BH_4-treated group, from prior to initiating treatment to when they had taken the drug for 2 or 5 years. In addition, growth impairment was also observed in patients on low-phenylalanine diets. In fact, individuals on long-term BH_4 treatment seemed to achieve similar developmental outcomes to those on more restricted diets. In conclusion, the author's study identified growth impairment in patients with phenylketonuria on BH_4 treatment, despite the fact that their natural protein intake increased.

Chapter 3 - Total nitrogen in grapes consists of inorganic and organic forms. Ammonia salts and amino acids are the most prevalent form of total nitrogen by weight in grapes. They constitute, except proline under good nitrogen content conditions, the assimilable nitrogen fraction, which is used by the yeast in the fermentation process.

Moreover, amino acids are also important because they are precursors of a larger number of compounds in grape and wine. Thus, the phenolic compounds are synthesized by the plant through the flavonoid pathway, starting with phenylalanine as a precursor.

This amino acid is also metabolised in the course of the fermentation by the yeasts, producing 2-phenylethanol and 2-phenylethyl acetate, which are important compounds for the wine aromatic quality.

Therefore, the foliar application of this amino acid in the vineyard could modify the phenolic and aromatic composition of grapes and wines. In recent years, some studies have been conducted about different treatments in vineyard that modified berry composition. These treatments have been carried out to improve the phenolic compound content and the aromatic quality of grapes and wines. However, the effect of the foliar application of phenylalanine on the composition of grape and wine has recently begun to be studied. For these reasons, this paper reviews the role of phenylalanine in the fields of viticulture and enology, focus on its influence on the phenolic and aromatic quality of grape and wine.

Chapter 4 - In order to investigate the penetration of amino acids and peptides into hair keratin fibers, cross-sectional samples of virgin white human hair treated amino acids and peptides were prepared. The authors have developed a new method for analyzing the penetration of amino acids and peptides into hair keratin fibers using optical microscopy and dyeing technique. The hydrophilic amino acids and peptides did not penetrate at all, while the hydrophobic amino acids, such as L-phenylalanine (Phe) penetrated into the virgin white human hair. In addition, the cross-sectional samples at various depths of virgin white human hair treated with Phe was directly analyzed without isolating the cuticle and cortex using Raman Spectroscopy. The Phe hydrophobic amino acid content at various depths of the virgin white human hair increased by performing the Phe treatment (at 50°C for 16 h), indicating that Phe deeply penetrated into the virgin human hair.

In: Phenylalanine
Editor: Darlene Williamson

ISBN: 978-1-63482-501-6

Chapter 1

BIOLOGICAL ROLES OF PHENYLALANINE AND ITS DERIVATIVES; THE GENETIC DISORDER PHENYLKETONURIA AND ITS POSSIBLE TREATMENT; POTENTIAL SEPARATION METHODS OF PHENYLALANINE ENANTIOMERS AND BIOLOGICAL FUNCTIONS OF D-PHENYLALANINE

Marek Graff
Department of Chemistry, Warsaw University, Poland

ABSTRACT

Phenylalanine (Phe) is an α-amino acid with the aromatic ring in molecule, similar to other amino acids containing aromatic rings, like tyrosine and tryptophan. Phenylalanine is classified as nonpolar because of the hydrophobic character of the benzyl group as the side chain. This amino acid is used as a precursor in a tyrosine synthesis and other compounds containing a six-membered aromatic ring, like dopamine, and compounds known as catecholamines: epinephrine (adrenaline), norepinephrine (noradrenaline), and melamine (the skin pigment). Phenylalanine was discovered in yellow lupine by Schulze and Barbieri in 1879, and synthesized by Erlenmeyer and Lipp three years later. In nature, phenylalanine is found in the breast milk of mammals. Excess of phenylalanine in the body is known as phenylketonuria (PKU) – it is a genetic disorder, related to lack of phenylalanine hydroxylase (PAH), an enzyme catalyzing the hydroxylation of phenylalanine to tyrosine (in aromatic ring). PKU is described as an autosomal recessive inborn error of metabolism. Test of the Phe level in blood is carried out for new born babies, for eliminating potential PKU; lack of control and possible of PKU can cause of brain damages. Patients with PKU have to control their intake of phenylalanine. This amino acid is also used in nutritional applications like aspartame – the artificial sweetener. From another hand, D-phenylalanine is found in proteins in small amounts, especially in processed ones, but biological roles of D-amino acids remain unclear. However, it is known the pharmacological activity of D-phenylalanine at the niacin receptor. Additionally, D-phenylalanine has also an antidepressant activity. It is known the fact that a cross the blood-brain barrier for D enantiomer is less efficient in comparison to L one. L-phenylalanine is synthesized on an industrial scale the bacterium *Escherichia coli* and used for medical, feed, and nutrition applications.

PHENYLKETONURIA: DESCRIPTION, ITS GENETIC ORIGIN AND PRIMARY MEDICAL TREATMENT

Phenylalanine (Phe) is coded for DNA and codons are UUU and UUC. Phenylketonuria (PKU) was first discovered by Folling (1934) in Norway. As mentioned above, phenylketonuria is the genetic (autosomal, recessive) inborn error of metabolism coming from lack of phenylalanine hydroxylase (PAH). Non-diagnosed and non-treated can cause hyperphenylalaninemia (HPA) and a neurotoxic effect an impaired postnatal cognitive development in consequence, according to Zurflüh et al. [1]. First introduction to the disease PKU was reported in 1949 at the University Children's Hospital Zurich by Fanconi and his co-worker Bickel. They performed a test using $FeCl_3$ in every retarded patient. Later, after moving to the University Children's Hospital in Birmingham, Bickel in cooperation with Gerrard and Hickmans carried out the same test in young patients in 1953 and observed hypothetical causal relationship between the Phe excess in the biological fluids and the human brain damage, which may be treated by reducing Phe intake. They used the diet with Phe-limited casein hydrolysate as the main protein source. However, initial results were tragic.

It is known that phenylketonuria occurs approx. in 1 in 10,000 births and the typical treatment for PKU is a strict dietary regimen [2]. Steinfeld et al. reported that in PKU patients receiving tetrahydrobiopterin (BH(4)), the tolerance increases towards a Phe intake. Probably this treatment may be used as an option in patients with mild PKU.

Newborn screening for PKU was elaborated by Guthrie and Susi in 1963 [3]. Their researches associated with cancer, became the inspiration to them in development of treatment for PKU: they observed clear relationship between a phenylalanine-restricted diet and blood Phe levels, which may be monitored. In their cancer investigation, they used variety of the bacterial test for screening of different substances in the blood of patients who were treated for cancer. These tests were based on the 'competitive inhibition'; compounds were removed, which could prevent growth of bacteria in culture plates (i. e. the inhibition process was stopped), when large Phe amounts was present in a blood spot that was added to the plate. They used filter paper discs soaked in serum from the patient and observed that total blood worked equally well and facilitated newborn screening. Authors carried out tests (400,000 infants in first two years of life) in 29 states and 39 cases of PKU were identified (an incidence of about 1 per 10,000). It is worth emphasizing that none was missed by screening. Treatment for PKU is recommended successfully by the National Association for Retarded Children in 37 states by 1967.

According to current knowledge, the phenylalanine hydroxylase (PAH) gene is localized on the chromosome 12 (NCBI Location Chromosome; Chr.12: 103232104 -103311381) [4]. Gibbs reported the disease PKU is caused for two mutations in *PAH* gene, first CM962547 (chr12:101761685-101761687) and the second CM941128 (chr12:101784521-101784523) [5]. Author compared human and macaque genes and found that for the macaque, deleterious alleles were identified: *OTC* (ornithine transcarbamylase) and *PAH*, which are associated in human with OTC deficiency and PKU. In addition, these mutations seriously perturb the typical amino acid levels in serum. Gibbs did not observe (for macaques) an increased phenylalanine or ammonia level, as a potential consequence of these changes, but for glycine slightly higher concentration was observed, and lower for cystine and cysteine, in comparison to human.

Patients with phenylketonuria, except mental dysfunction, have symptoms like: epilepsy, peculiarities of gait, stance, and sitting posture; 'mousy' odor; light pigmentation and eczema (Paine, 1957). However, some authors doubt that phenylketonuria may cause cataracts or syphilis: Pitt and O'Day (1991) in a group of 46 patients with untreated PKU found only 3 ones with cataracts, and Levy et al. (1970) found patients with phenylketonuria and syphilis in common less than 1%.

Zurflüh et al. argued BH(4)-responsive HPA as an option of lack of PAH, which was affected by a specific mutations in the *PAH* gene (they identified 57 mutations). Authors suggested that the expected average occurrence of BH(4)-responsiveness in European and Far East populations was calculated to be 55-57%, by using Hardy-Weinberg formula. They found that three most often found mutations are p.A403V, p.R261Q, and p.Y414C, for >5% of BH(4)-responsive patients, but for mutations: L48S, p.I65 T, p.R158Q, p.R261Q, and p.Y414C results were incoherent.

Blau et al. reported that in about 20% of PKU patients, BH(4) may increase the PAH activity [6]. Additionally, these patients may have specific dietary freedom, when BH(4) is added to their Phe-restricted diet. Potential treatments should include an enzyme substitution with phenylalanine ammonia lyase, degrading Phe, plus a gene therapy for restoring the PAH activity.

According to Kaufman et al., who analyzed liver biopsy samples from patients with HPA and their families, a level of PAH in patients with HPA was about 5% of normal and 7.3-10% of normal in their parents [7]. Authors observed the presence of the negative interallelic complementation, which includes protein-protein interactions between subunits in a mutlimeric enzyme.

Burgard et al. (1996) observed an enzyme activity of *in vitro* residual PAH in patients who had PKU and found that patients with the PAH activity greater than 20% had mild PKU, and patients with the PAH activity lower than 20% had classical PKU. Additionally, the R261Q mutation with the significant amount residual enzyme activity, producing higher Phe amounts than needed, hypothetically can be described as the negative intraallelic complementation effect as the reason for higher than needed diagnostic Phe levels.

Kalaydjieva et al. found, using a direct sequencing technique of the *PAH* gene, silent mutations in codons 232, 245, and 385, linked to specific Restriction Fragments Length Polymorphism haplotypes in several Caucasian populations [8]. All three mutations generate a new restriction site and can be simply detected by PCR. Authors argued the usefulness of the silent mutations for potential treatment is determined by the haplotype distribution in the target population.

According to current knowledge, $\geq$ 50% patients with PKU have a mild clinical phenotype. Muntau et al. studied a therapeutic efficacy of BH(4) in the treatment of mild phenylketonuria [9]. HPA, coming from PAH deficiency, is associated with this genetic metabolic disease. In treated patients with PKU, a low-Phe diet is used, and alternative treatments are rather ineffective. Authors used, in 38 children with PAH deficiency, a combined phenylalanine-tetrahydrobiopterin loading test and analyzed the *in vivo* rates of the [^{13}C]Phe oxidation process. They observed in 87% patients with mild HPA or mild PKU, a lowered blood Phe level by BH(4). Muntau et al. identified 7 mutations (P314S, Y417H, V177M, V245A, A300S, E390G and IVS4-5C->G) as possibly associated with responsiveness to BH(4), 6 mutations (A403V, F39L, D415N, S310Y, R158Q, and I65T) as potentially associated, and 4 mutations (Y414C, L48S, R261Q, and I65V) as rather unassociated with this phenotype. Authors suggest that in patients with mild HPA phenotypes, the BH(4) responsiveness is common. However, responsiveness cannot be regularly expected on the basis of genotype, especially in compound heterozygotes.

Matalon et al. investigated the BH(4) responsiveness to PAH deficiency in patients with atypical and classical PKU [10]. They performed a *PAH* gene mutation analysis by using denaturing gradient gel electrophoresis and a gene sequencing technique. Patients with classical, atypical, or mild PKU, took orally BH4 10 mg/kg, and blood phenylalanine and tyrosine levels were measured using coupled MS/MS (tandem mass spectrometry) at 0, 4, 8, and 24 hours intervals. 21 patients (58%) had decreased Phe level in blood – 12 patients with classical, 7 atypical and 2 with mild PKU. In 15 patients (14 with classical and 1 with atypical) no respond to the BH(4) challenge was observed. Matalon et al. suggested, according to mapping the mutations that responded to BH(4) on the PAH enzyme, mutations were found in catalytic, regulatory, oligomerization, and BH(4) binding domains. Additionally, 5 new mutations were also found. In conclusion, authors suggest higher than expected the number of PKU mutations respond to BH4, and these mutations are present on all the domains of PAH.

MEDICAL TREATMENT OF PKU

PKU treatment involves a low-Phe diet, as mentioned above. Additionally, a neonatal screening method was introduced 30 years ago, what could significantly decrease the number of neurological dysfunctions associated with this disorder [11]. Thompson et al. observed strong white matter abnormalities direct associated with blood Phe levels > 15 mg/l. In 34 treated patients with PKU and the low Phe diet, they observed routine neonatal screening for 25 ones, but in 9 were missed in the screening programme. In addition, they found in two patients abnormalities using MRI (magnetic resonance imaging).

Griffiths et al. investigated intelligence quotient (IQ) scores of 57 selected British children at age 8 with early-treated normal PKU, using some variants of the 'Wechsler intelligence scale for children' (WISC) [12]. Authors analyzed relationships between WISC and dietary parameters like serum Phe levels and socioeconomic factors. They argued that in treatment of PKU patients, early and continuous therapy did not automatically lead to stabilization of overall IQ, and verbal intelligence did not lead to normalization, but spatial intelligence remained poor; for children in primary school: ones with annual median Phe concentration at 360 μmol/l (suggested as upper limit) had higher IQ at 10 points in comparison to other children.

Weglage et al. investigated patients with PKU at aged 10-18 years and diabetes (two groups, 42 patients in each group), matched for age, socioeconomic status and sex, and compared the results with healthy controls (n = 2900) from an epidemiologic study [13]. They analyzed the blood Phe and HbA1c (glycated hemoglobin) concentrations, using IQ test and Child Behavior Check List, according to Achenbach. Authors found that internalizing problems (depression, physical complaints or social isolation, nervousness) were considerably higher for phenylketonuric and diabetic patients, but externalizing problems were not observed. In addition, no relationships were found between the psychological profile and the biochemical parameters, the IQ, or the socioeconomic status of phenylketonuric and diabetic ones.

Levy et al. investigated plasma Phe levels in pregnant women with classical PKU, which were not under metabolic control during minimum the first 2 trimesters of pregnancy (c_{Phe} = 1260 μmol/l) and pregnant women with classic PKU under metabolic control (c_{Phe} = 360 μmol/l) [14]. They observed for higher plasma Phe levels (mothers without metabolic control during their pregnancies), a tendency for increasing the hypoplasia of the corpus callosum presence (they used brain magnetic resonance imaging).

Rouse et al. investigated phenylketonuria in untreated women at pregnancy and maternal PKU offspring [15]. They analyzed relationships between high Phe plasma levels and the malformations of offspring incidence like microcephaly, congenital heart disease (CHD), mental retardation, and intrauterine and postnatal growth retardation, early abnormal urologic signs and other major and minor defects. The pregnant women were divided according to their Phe plasma levels during critical gestational periods and typical for Phe exposure throughout pregnancy: ≤ 360, 361-600, 601-900, ≥ 900 [μmol/l]. They observed the correlation between increasing of congenital abnormalities and higher as typical maternal Phe levels: summary, 14% of infants had CHD (none in the lowest range of the maternal Phe levels). In addition, at the lowest maternal Phe level (≥ 360 μmol/l), only 6% infants had microcephaly, 4% were with postnatal growth and 0% with intrauterine growth retardation. For comparison, at the highest Phe level (≥ 900 μmol/l) the frequencies of these diseases were 85%, 51%, and 26%, respectively. Authors argued that pregnant women have to be on the low-phenylalanine diet to achieve Phe level at ≥ 360 μmol/l prior to beginning and should maintain throughout pregnancy.

Waisbren et al. investigated a risk of untreated maternal PKU and developmental problems in offspring [16]. They observed, for offspring of women with metabolic control prior to their pregnancy, score of 99 (Scores on the McCarthy General Cognitive Index, GCI). For comparison, for women without metabolic control by 20 weeks, GCI score was 2 for 47% women. In addition, 30% of children born to mothers with phenylketonuria had behavioral and social problems.

Rouse et al. studied a relationship between a Phe level and microcephaly, CHD, and development of their offspring [17]. They analyzed the blood Phe level, sonograms in pregnant women (at 18 to 20 and 32 weeks' gestation), and Wechsler Adult Intelligence Scales and DNA tests were carried out. These tests were helpful in identification of the mutations associated with PKU. For offspring, Bayley Mental Developmental Index and Psychomotor Developmental Index tests were done during first and second year of life. Authors observed increasing (3-fold risk) of having microcephaly in infants with congenital heart defect, which mothers had untreated PKU at pregnancy. If pregnant women with PKU were treated, none of them had offspring with CHD. However, Rouse et al. did not find any direct relationship between PKU and the specific *PAH* mutation.

Levy et al. investigated offspring from maternal PKU pregnancy and offspring from control pregnancies, and observed congenital heart disease in the offspring (2%) from mothers without PKU, but with HPA [18]. According to Levy et al., if a Phe level is higher than 1800 μmol/l in the mother blood, a risk for bearing a child with Coronary Heart Disease (CHD), significantly increases (the risk threshold of CHD is expected at 900 μmol/l). In addition, for offspring with CHD, coarctation of the aorta and hypoplastic left heart syndrome were observed at the higher level than expected – in comparison to patients with CHD in the general population (patients, whose mothers had no PKU).

Brumm et al. investigated psychological and psychiatric problems of patients with PKU [19]. They found in untreated PKU much more than expected attentional problems of early-treated children and adolescents, like autistic features, decreased achievement motivation, hyperactivity, school problems, aggression, lowered social competence or autonomy, and low-self-esteem. In addition, in early adulthood time, early-treated patients may turn into lack of autonomy and low self-esteem, but also are possible symptoms like higher nervousness, depressed mood, phobias, lowered positive emotions, social isolation and maturity deficits. Authors observed that if children with PKU were treated, behavioral problems could be reduced, and tendencies like to be less confident and happy, were present. In adults with early treated PKU, symptoms like social introversion, and anxiety-related disorders were comparable with patients without PKU, but rates of depression were higher. They suggested an existence of a biological basis of psychiatric dysfunction, but none PKU-specific psychiatric 'phenotype' had been found: not every patient with PKU had similar behavioral symptoms as described above.

Scriver et al. investigated the Phe concentrations in 10 patients, ages 26-54 years (5 males and 5 females), under two dietary protocols and four times a day [20]. They argued that the typical phenylalanine concentrations in blood are:

- adults: 58±15 μmol/l,
- teenagers: 60±13 μmol/l,
- children: 62±18 μmol/l,
- newborn (babies): 120 μmol/l (upper limit),
- patients with untreated classical PKU: 2400 μmol/l.

van Spronsen et al. studied a rate of conversion of phenylalanine into tyrosine (Tyr) and potential consequences for tolerance to dietary Phe and as a result the clinical severity of PKU [21]. It is known that in PKU patients, the correlation is observed: a decreased conversion rate of Phe into Tyr is associated with the higher tolerance for Phe. van Spronsen et al. investigated this correlation using an *in vivo* stable isotope study, similar technique like Thompson and co-workers (normal *in vivo* hydroxylation rates of Phe in every PKU patient) [22-24]. Authors observed very low (0.13-0.95 μmol/kg/h) and normal (4.11 and 6.33 μmol/kg/h) conversion rates in patients and controls, respectively. No relationship between the *in vivo* hydroxylation rates and the tolerances was found.

Hanley et al. investigated disabilities in treated patients with PKU, like decreased production of neurotransmitters as a consequence of a poor Tyr transport process across the neuronal cell membrane (this symptom is known as hypotyrosinemia) [25]. They found non fasting plasma Tyr level:

- 41.1 μmol/l in 99 classic PKU patients,
- 53.3 μmol/l in 26 mild (atypical) PKU patients,
- 66.6 μmol/l in 35 non-PKU mild HPA patients.

For comparison, non-fasting plasma Tyr levels for patients without hypotyrosinemia, were:

- 64.0 μmol/l in 102 non-PKU patients (their hospital biochemistry database),
- 69.1 μmol/l in 58 volunteers (private office practice),
- 64.0-78.8 μmol/l in infants, children, and adolescents (literature review).

Leuzzi et al. investigated the brain Phe concentrations in 10 off-diet PKU patients, using *in vivo* magnetic resonance spectroscopy (^{1}H MRS) [26]. They detected a brain MRS signal (resonating at 7.36 ppm) coming from the presence of creatine + phosphocreatine. They observed an abnormal concentration of brain Phe in all patients with PKU (ranging from 0.030 to 0.074 μmol/l), associated with wide scale of plasma Phe (ranging from 724 to 2800 μmol/l). Authors also found white-matter alterations in all patients, but no correlation was observed between white-matter alterations and the brain Phe concentrations. In addition, they argued that the brain Phe concentrations were correlated with clinical phenotypes more regular in comparison to plasma Phe.

According to Hanley et al. (1996) and Cipcic-Schmidt et al. (1996), for protection to the fetus and later intellectual development, an adequate and early diet is strong recommend during pregnancy. Brenton and Lilburn (1996) reported that 39 treated pregnant women with PKU, 2 of these offspring died of congenital heart disease and 4 needed surgeries for coarctation. No heart defects were observed in 34 pregnant women and their offspring, when diet was controlled (maternal phenylalanine levels; three times weekly).

Weglage et al. studied a (mild) neuropsychological impairment in early-treated phenylketonuric patients [27]. They observed 20 early-treated (9-30 days) adolescent phenylketonurics (10 male, 10 female) and 20 healthy control patients. Authors analyzed results of testing of: IQ (Culture Fair Intelligence Test-Scale 2; CFT-20), fine motor abilities (Motor Performance Task), sustained (Test d2) and selective attention (Stroop-Task). They observed the correlation between the high blood Phe level and neuropsychological deficits in PKU patients. Additionally, authors found that clinical-neurological status of treated phenylketonurics and controls were comparable. However, the lowered vulnerability of patients with PKU as a function of blood phenylalanine levels and potential neuropsychological deficits on ageing was significant.

Koch et al. investigated 125 children with long- and short-term treatment of PKU and a low-Phe diet [28]. They used magnetic resonance imaging and spectroscopy (MRI/MRS) to analysis of the brain Phe concentrations, and for the higher brain Phe concentrations, anomalous MRI spectra were detected. They observed in 41% of those off diet and 22% of continuers, diseases like eczema, asthma, headache, hyperactivity and hypoactivity and mental problems, including depression and phobias. In treated patients with a Phe-restricted diet, only two ones required psychiatric care (transient depression treatment).

Stegink et al. investigated an excessive use of aspartame (APM) (N-L-α-aspartyl-L-phenylalanine methyl ester; Figure 1) for the increasing plasma APM and Phe concentrations in individuals heterozygous for phenylketonuria (PKUH) [29]. In first study, PKUH patients (three males; three females) ingested 3 successive 12-oz servings of beverage at 2-ounce intervals. To one arm, aspartame was not added, to others – was 10 mg APM/kg per serving. Authors observed – for patients who ingested APM, increasing of plasma Phe level from 2.3 to 4.1 μmol/l above baseline value 30 to 45 min after every dose, but authors did not detect significantly the increasing of the plasma APM concentration. In addition, after repeated APM dosing (13.9 ± 2.15 μmol/l), the detected PKUH level (12.6 ± 2.11 μmol/l) was slightly, but not significantly higher, above the typical postprandial range. In second study, also 6 patients with PKUH ingested beverage (as a single bolus) providing 30 mg APM/kg. Stegink et al. observed the higher plasma Phe concentration when APM was ingested as a single bolus in comparison to that, when APM was ingested as a divided dose.

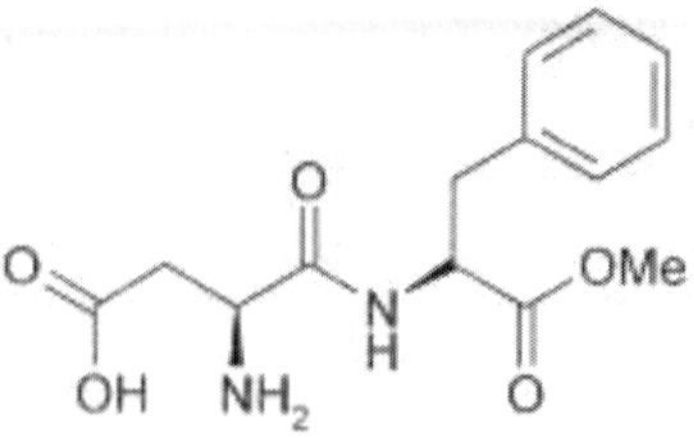

Figure 1. Chemical structure of aspartame.

Glushakov et al. studied effects of L-Phe on NMDA (N-methyl-D-aspartate)-activated currents (I_{NMDA}) in cultured hippocampal neurons from newborn rats by using the patch-clamp technique [30]. According to them, hippocampal N-methyl-D-aspartate receptors (NMDARs) are probably engaged in the regulation of memory formation and process of learning. In addition, the mechanism causing mental retardation in patients with PKU (≥1.2 mmol/l) remains unknown. L-Phe may reversibly and specifically decrease I_{NMDA} in a concentration-dependent manner (IC_{50} = 1.71 ± 0.24 mmol/l). However, authors did not observe that L-tyrosine could widely change I_{NMDA} (tyrosine is synthesized from L-Phe). Additionally, authors did not find any relationship between the L-Phe-I_{NMDA} concentration-response and the concentration of NMDA. They suggest that L-Phe may play role of a specific inhibitor of NMDAR current in hippocampal neurons by opposing the glycine-binding. In consequence, excess of L-Phe (causing PKU), plays the crucial role for the impaired NMDAR function in the development of mental retardation during PKU and the essential role for NMDARs in the learning and memory formation could be possible.

Glushakov et al. studied the L-Phe concentrations in a brain depresses glutamatergic synaptic transmission of mice with PKU [31]. They observed that during this process the GABA (gamma-aminobutyric acid) receptor activity in cultured mice neurons was not affected. In addition, if these depressant effects of L-Phe in the brain of mice with PKU are present, potential changes in postsynaptic glutamate receptors could be detected, because chronic impairment of the glutamate system may cause impaired brain function. They used a combination of liquid chromatography-mass spectrometry, patch-clamp, radioligand binding and western blot approaches; dissected forebrain tissue from homozygous and heterozygous PKU Pah^{enu2} mice. They observed circa six-fold greater in PKU mice (863.12 ± 17.96 mmol/kg) in comparison to their heterozygous counterparts (149.32 ± 10.23 mmol/kg). In addition, experiments with the receptor binding technique and [^{3}H]MK-801 showed noteworthy up-regulation of NMDA receptor density in PKU mice. Glushakov et al. reported that the expressions of glutamate receptors, like NR2A and (RS)-amino-3-hydroxy-5-methyl-4-isoxazolepropionic acid (AMPA) receptor Glu1 and Glu2/3 subunits, were increased, and the expression of the NR2B subunit was decreased. For the GABA α1 subunit, no change in expression was observed. Authors argued that in the brain functioning and potential disorders in patients with PKU, crucial roles may play glutamatergic synaptic transmission.

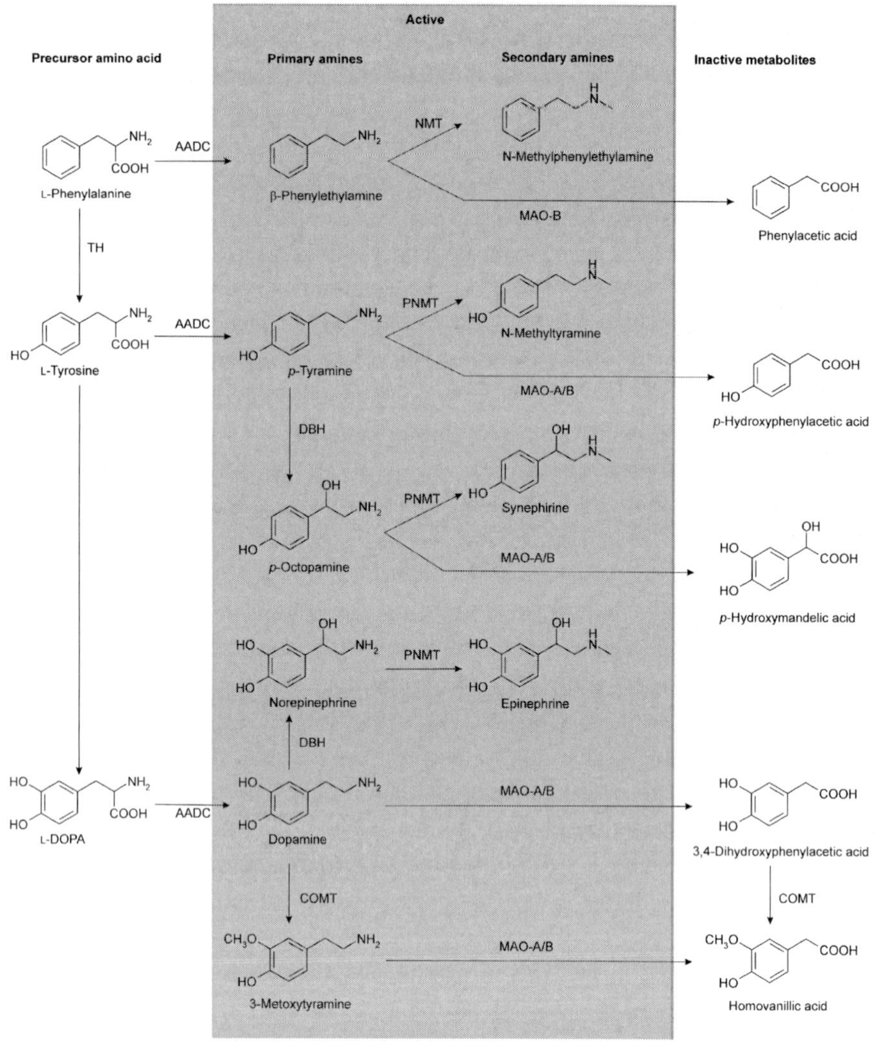

Figure 2. The main routes of TAs metabolism in vertebrates.

Derivatives of Phenylalanine and Their Biological Function

Phenylalanine is metabolized into a number of compounds, like tyramine, β-phenylethylamine (β-PEA), tryptamine and octopamine, which are biologically active amines, found in the body in trace amounts (Figure 2) [32]. These chemical species are naturally occurring amines, found in the diet, bacteria, fungi and in herbal plants [33]. In addition, ephedrine and cathinone – synthetic analogues may be also classified to them. Tyramine and β-phenylethylamine are found in red wine, fermented foods such as sausages and cheeses, chocolate and other food preparations containing cocoa [34]. Trace amines (TAs) are present in the mammalian brain at nanomolar levels [35-37]. Additionally, to this group may also be classified amphetamine and its analogues [38-40], for instance methylenedioxy, methamphetamine / MDMA ('ecstasy'), phenylpropanolamine and pseudoephedrine, synthetic proprietary sympathomimetic agents (Figure 3). These substances can cause vasoconstriction and a rise in blood pressure in human vascular system. Additionally, many researchers accept performing of these chemical species as indirectly acting sympathomimetic amines. Answer of amines is related to the release of noradrenaline from sympathetic neurons. Potential action of these amines may also be referred as direct vascular effects independent of a noradrenergic mechanism. Last papers reported that group of unique trace amine associated receptors (TAARs) were found in the brain and peripheral tissues (including blood vessels). The receptors were cloned and authors observed that TAs can bind to these receptors. They proposed that vasoconstrictor effects may be assigned as a part of this mechanism. Additionally, amphetamines and TAs can interact with TAARs (localized in blood vessels), and the vasoconstriction may be associated with the activation process of these receptors. This model can also explain cardiovascular toxicity of amphetamines. Drugs with increasing vascular actions, like tyramine, were described at first by Barger and Dale in 1910, but many aspects of this action still remains unclear. Amphetamines and its derivatives are drugs predominantly release neurotransmitters, in this case predominantly catecholamines, by a non-exocytic way, according to Sulzer et al. [41]. In many cultures, these drugs play a crucial role in medicine, as mental and behavioral stimulants; however, they can create neurodegeneration and addiction. Sulzer et al. also proposed two mechanisms of the redistribution of catecholamines to the cytosol from synaptic vesicles, and induction of inversed transport of transmitter *via* plasma membrane uptake carriers. In addition, other effects are also observed; for instance, affect extracellular catecholamine levels, like uptake inhibition, effects on exocytosis, neurotransmitter synthesis, and metabolism. Amphetamine and its derivatives may be considered orchids of the psychoactive drugs. Their actions are close and results are comparable – they do changes in dopamine release in the striatum, elimination rates, and have similar pharmacokinetic properties (Melega et al., 1995). Lamb and Henningfield (1994) argued that equal doses of AMPH and METH are not distinguished in the human discrimination analysis. Some authors reported (Shoblock et al., 2003), in rodents, AMPH can elicit rather more locomotor activity in comparison to METH. In addition, AMPH is more addictive, favored by drug addicts and stronger than METH psychostimulant for human.

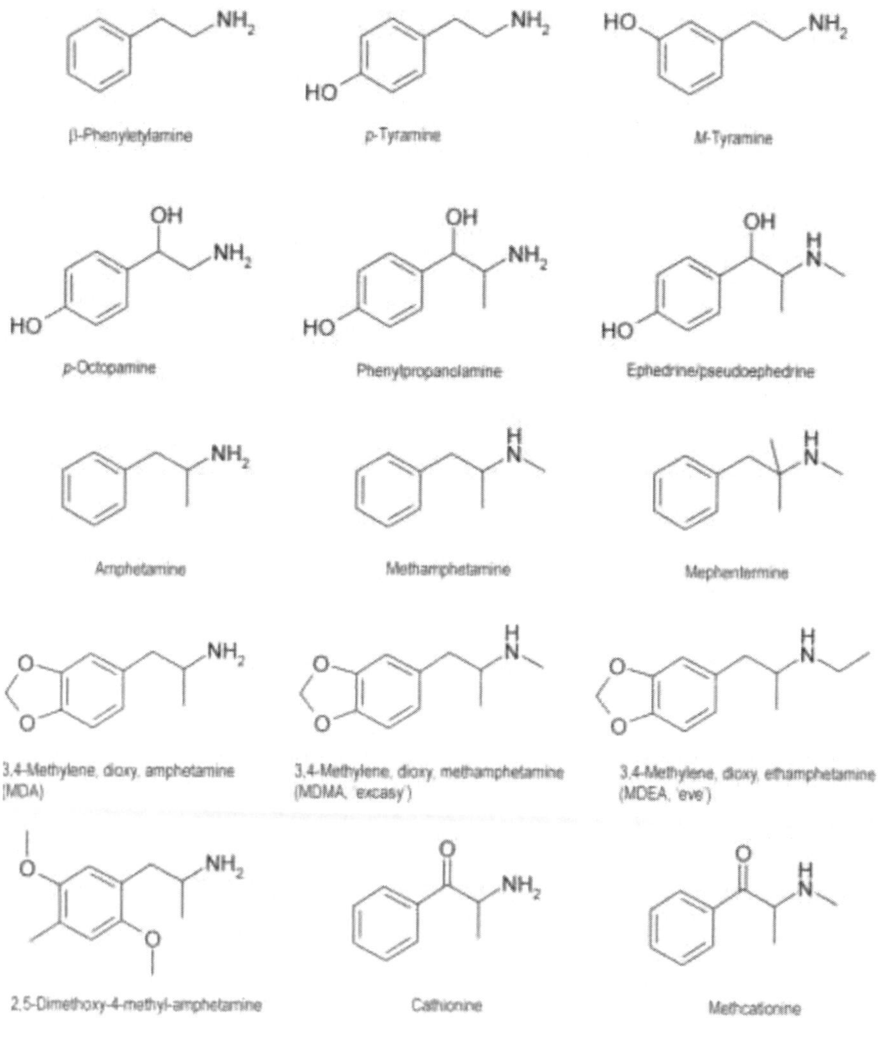

Figure 3. Chemical structures of TAs, amphetamines and sympathomimetic amines.

According to Burchet and Hicks, TAs isomers are synthesized in mammalian brain and peripheral nervous tissues [42]. Additionally, these compounds are involved widely in human disorders of affect and cognition. They are present at very low concentration and heterogenic distribution in the mammalian brain and their metabolism have high rate. It is suggested that TAs play the important roles in the coordination of biogenic amine-based synaptic physiology. If the concentrations of these amines are high, the presynaptic 'amphetamine-like' effects on catecholamine and indolamine release may be observed. If concentrations are low, postsynaptic modulatory effects are present and increasing the activity of other neurotransmitters, especially serotonin and dopamine, is observed. The authors reported that electrophysiological effects of TAs are opposite in comparison to these neurotransmitters, and they suggest that the TAs can interact with selective receptors.

Lindemann and Hoener studied a physiological function of biogenic amine neurotransmitters, found in the mammalian nervous system in trace amounts [43]. According to them, a novel family of G protein-coupled receptors (GPCRs), which were discovered recently, probably GPCRs are highly specific for TAs and indicate a potential role for TAs as vertebrate neurotransmitters and neuromodulators. However, it is considered that activations of the majority of GPCRs by the TAs are not realized. For one hand, main human illness: depression and schizophrenia have been studied for many years, but an action at molecular level (interactions of amines with specialized receptors) remains incompletely understood for today. Clearing this action can help in drug development (molecular pharmacology) in therapy of these neurological diseases.

Frascarelli et al. investigated cardiac effects of TAs like *p*-tyramine, β-phenylethylamine, octopamine and tryptamine [44]. These compounds may couple with GPCR receptors. Authors made a perfusion (a liquid poured through or over an organ or tissue) of isolated rat hearts in the presence of the TAs, observing the hemodynamic variables. They used radioligands like [^{3}H]-*p*-tyramine and [^{125}I]-3-iodothyronamine in rat ventricular tissues. According to authors, octopamine, β-phenylethylamine and tryptamine caused a dose-dependent negative inotropic effect and reduced cardiac output was observed: IC_{50}=109 μmol/l, 159 μmol/l, and 242 μmol/l, correspondingly (IC – inhibitory concentration). In addition, for thyronamine and 3-iodothyronamine similar effect was detected: IC_{50}=94 μmol/l and 27 μmol/l, respectively. Authors confirmed the negative inotropic effect of octopamine in the papillary muscle preparation and suggest that except tryptamine, all TAs may increase the heart rate. In addition, they observed increasing the negative inotropic effect of TAs by the tyrosine kinase inhibitor genistein (a phytoestrogen found in soy, having antioxidant and anticancer properties). They also observed differences in binding of [^{3}H]-*p*-tyramine and [^{125}I]-3-iodothyronamine in ventricular tissue: [^{125}I]-3-iodothyronamine was not displaced by *p*-tyramine, but [^{3}H]-*p*-tyramine was – by 3-iodothyronamine. They concluded that the TAs and thyronamines may cause the negative inotropic effects, in rank of potency: 3-iodothyronamine > thyronamine = octopamine = β-phenylethylamine >> tryptamine = *p*-tyramine.

BIOLOGICAL FUNCTION OF D-PHENYLALANINE

D-phenylalanine has a secondary biological and physiological role – it is reported that D-phenylalanine may be a potential drug in a PKU therapy, and D–phenylalanine derivatives have a potential chance to be naturally-occurring sweeteners. It is also suggested that D-phenylalanine residues may be used for synthesis of new drugs (for instance, peptides).

Singh et al. investigated interactions of L-Phe self-assembles to amyloid fibrils in many conditions and the transformation to a gel state at saturated concentration [45]. They used techniques like NMR, SEM, light scattering intensity (as time function), and dynamic light scattering (as function of temperature and ionic strength). The aim of the research was the detailed study of self-assembly of Phe interaction in amyloid formation and their potential toxicity in PKU. They observed the higher fluorescence intensity (for 800 s – three times higher) for DL-Phe in comparison to every enantiomer separately. Self-assembly kinetics of enantiomeric and racemic Phe was monitored by the thioflavin binding fluorescence assay. They argued that L-phenylalanine molecules may interact together mostly by hydrophobic interactions (π-π stacking) in the fibril formation process, but also hydrogen bonding creation were observed between symmetrical –C=O··HOOC– and electrostatic interaction between $–NH_3^+$ $^{-}$OOC-groups [46,47]. Addition of D-phenylalanine before the fibrils formation may stop the process and their increasing to flakes (the expansion of fibers is blocked). Singh et al. suggest that D-phenylalanine may be the modulator of L-phenylalanine amyloid formation, and it is potential therapeutic molecule in PKU.

Christianson et al. studied the structures of the complexes of carboxypeptidase A with the amino acids D– and L-phenylalanine and D-tyrosine by using crystallographic methods, including x-ray to a resolution of

2.0 Å [48]. They observed that amino acids bind to the enzyme in the COOH-terminal hydrophobic part: the carboxyl group of amino acids can interact with Arg-145 and amino group with Glu-270. Additionally, interaction between the α-amino groups of D-Phe and Glu-270 is stronger in comparison to L-Phe.

Dhaneshwar et al. studied a mutual action of amide prodrugs synthesized from 4-aminosalicylic acid with D-phenylalanine and L-tryptophan in the therapy of inflammatory bowel disease [49]. They observed that analyzed prodrugs were stable in aqueous buffers (pH 1.2 and 7.4) over 10 h. In addition, the release of 4-aminosalicylic acid from the prodrugs (in rat fecal) was in the range of 86-91% over a period of 20 h (half-lives were equal 5 h 43 min and 3 h 32 min, respectively). They argued that synthesized prodrugs containing 4- and 5-aminosalicylic acids, can be less toxic in comparison to sulfasalazine, but the pharmacological action and advantages are similar to sulfasalazine.

Masuda et al. investigated sweet taste receptors, which are classified as GPCR ones and usually form heterodimeric structures with T1R2 and T1R3 subunits [50]. It is known that D amino acids like as D-phenylalanine and D-tryptophan are naturally-occurring sweeteners. In addition, D-phenylalanine derivatives with substituents like the trifluoromethyldiazirinyl moiety at 3- or 4-position of phenylalanine, have a potential chance to use as sweeteners. According to authors, each subunit contains a big amino-terminal domain (it can link *via* a cysteine-rich domain) and it is localized at an extracellular site to a seven transmembrane helical domain. For responding to sweet taste of wide spectrum of chemical substances including naturally-occurring carbohydrates, glycosides, D-amino acids, and synthetic chemical species like aspartame, saccharin and sucralose, it is responsible in human the heterodimeric sweet taste receptor (hT1R2-hT1R3). Though different chemical structure of these sweeteners, all mentioned compounds interact with the same sweet taste receptor. Masuda et al. observed that activation of the sweet taste receptor was easier for D-amino acid derivatives in comparison to L ones.

Zhang et al. studied the enantioselective binding of L-/D-phenylalanine (optionally) to calf thymus DNA [51]. They used techniques like absorption, circular dichroism, fluorescence quenching, viscosity, salt effect and emission spectroscopy. They observed no difference in the emission spectra of L/D-Phe (60.5 mol/l) in the presence of ctDNA (1→8: 0, 30.3, 60.6, 90.9, 121.2, 151.5, 181.8, 212.2 mol/l), but they found small differences in the fluorescence quenching process of ctDNA linked optionally with L-phenylalanine and D-phenylalanine: for L enantiomer the fluorescence (F_0/F) was higher. According to that, they argued that Phe binding to DNA was enantioselective.

Dhaneshwar et al. studied an aceclofenac' pharmacological action, nonsteroidal anti-inflammatory drug used in arthritis therapy, because it relieves inflammation and associated pain [52]. However, aceclofenac may induce erosion and ulcers in the gastrointestinal tract. Authors synthesized a mutual prodrug of aceclofenac, using as a gastro-protective carrier system the D-phenylalanine for decreasing the ulcerogenic potential and increasing efficiency of aceclofenac as an anti-arthritic agent. Their research showed that D-Phe may enhance the complete pharmacological and ulcerogenic profile of aceclofenac. In addition, they demonstrated its utility as an effective gastro-sparing chemical delivery system. They also observed that the conjugation of aceclofenac with D-phenylalanine also may widely increase its analgesic, anti-inflammatory and anti-arthritic activities.

Lovelock et al. studied the function phenylalanine ammonia lyases (PALs) from a family of 4-methylideneimidazole-5-one (MIO) cofactor dependent enzymes, converting (in prokaryotic and eukaryotic organisms) L-phenylalanine into trans-cinnamic acid [53]. They observed also non-stereoselective reaction of MIO with L- and D-phenylalanine derivatives. However, for D-enantiomer, if mutagenesis of key active site residues is occurred, D-amino acids are synthetized, instead estimated L-amino acids in amination reactions catalyzed by bacterial and eukaryotic.

Solanas et al. investigated chemical species like the cationic antimicrobial peptide gramicidin S, cyclo(Val-Orn-Leu-D-Phe-Pro)$_2$, in which D-Phe residues were changed by different surrogates, like D-Tic (1,2,3,4-tetrahydroiso-quinoline-3-carboxylic acid) [54]. According to NMR spectra, the distance between the D-Tic aromatic ring and the Orn δ-amino group was decreased. Authors observed slight increased therapeutic index for new compounds, because of a higher drop in the hemolytic effect.

Zhang et al. investigated a sensitive and highly enantioselective immunosensor, which may be used for the detection of chiral amino acids and a potentiostatic step method as an analytical technique (authors measured capacitance) [55]. They attached the compounds (hapten 4-amino-D-phenylalanine) to the surface of a gold electrode *via* mercaptoacetic acid and used the potentiostatic step method for detection. They reported that detection limit of 5 pg of antibody/mL was achieved.

SEPARATION OF ENANTIOMERS FROM PHENYLALANINE RACEMIC MIXTURE

For enantiomers separation of amino acids, including phenylalanine, mostly an enantioseparation process is used, for instance, membranes like chiral thermo-sensitive or channel type, and solution-diffusion. In others words, a stationary phase in column chromatography contains mostly chiral polymers, DNA and others.

Khan et al. investigated of Phe racemic mixture using imprinted microbeads synthetized by a novel modified suspension polymerization method without derivatization of the water-soluble template molecule [56]. According to authors, first successful enantioseparation of phenylalanine was carried out by using D-Phe imprinted P(MAA-co-EGDMA) microbeads as HPLC stationary phase. They used to the separation of the Phe enantiomers, a stainless steel column filled by D-Phe imprinted P(MAA-co-EGDMA) microbeads (column backpressure was less than 300 psi), and a mobile phase as an organic-aqueous buffer solution (9-18% (v/v): EtOH in a 0.030 mol/l acetate buffer solution. They detected the separation factor more than 2.56, with the resolution of 1.38.

Feng et al. reported that a separation of phenylalanine racemic mixture is possible by using a novel chiral thermo-sensitive membrane and phase inversion method with a blend of chiral micro-gels and PVDF-g-PNIPA [57]. This compound – PVDF-g-PNIPA – poly(vinylidene fluoride)-grafting-poly(N-isopropylacrylamide), a thermo-sensitive polymer, was synthetized *via* a radical copolymerization process from poly(vinylidene fluoride) (PVDF) backbone and poly(N-isopropylacrylamide) (PNIPA) side chains. Authors prepared chiral micro-gels from a chiral monomer and N-isopropylacrylamide (NIPA) by reaction of acryloyl chloride and L-phenylalanine. They observed, for water flux of the membrane, the sensitivity to temperature and the permeability of D-Phe was higher in comparison to L-phenylalanine. In addition, a larger percentage of enantiomeric excess (e.e.%) at 25°C was observed for the chiral thermo-sensitive membrane *versus* the chiral membrane at 40°C. They described the analyzed model of membrane as the solution-diffusion one.

Matsuoka et al. studied a chiral separation of DL-Phe using DNA-immobilized chitosan membranes having various pore sizes [58]. They observed that the chitosan membranes covalently bound six times more DNA in comparison to the cellulose membranes. DNA-immobilized chitosan membranes with a pore size <6.4 nm were used for effectively permeation of D-Phe. For comparison, L-Phe was also adsorbed on the DNA-immobilized chitosan membranes with a pore size <6.4 nm and authors observed a little difference in the adsorption of every enantiomer on the membrane. In addition, they argue that DNA has a higher binding affinity (equilibrium binding constant) for L-phenylalanine in comparison to D-enantiomer. According to their previous researches, the preferential permeation through DNA-immobilized cellulose membranes for D enantiomer may be carried out with a pore size <2.0 nm (MWCO< 5000; molecular weight cut-off) and for L enantiomer – with a pore size >2.0 nm (MWCO> 5000) [59]. In addition, the chiral separation of the racemic Phe solution was more effective when amount of NH_2 residues in the DNA-immobilized cellulose membranes was higher. Contribution of NH_2 residues may increase in the amount of bound Pt complex existing on the membranes, and the amount of immobilized DNA is greater in consequence [60]. They classified the DNA-immobilized chitosan membranes as the channel type membranes.

According to Trost and Fischer, D-phenylalanine and D-methionine are biotransformed to corresponding α-keto acids by D-amino acid oxidase (DAAO, EC 1.4.3.3) [61]. By-product in the DAAO-catalyzed reaction is hydrogen peroxide. For removing of H_2O_2 generated during the biotransformations, authors tested three different commercially available catalases: bovine liver, *Aspergillus niger* and *Micrococcus lysodeikticus*. According to Trost and Fischer, the most effective was *Micrococcus lysodeikticus*. The comparison of reaction yields carried out with and without catalase presence were 8.26 g l^{-1} h^{-1} and 0.14 g l^{-1} h^{-1}, respectively. In addition, D-amino acids are transformed in α-imino acids and finally in α-keto acids.

Ul-Haq et al. investigated a separation of Phe racemic mixture using enantioselective poly[(acrylonitrile)-co-(acrylic acid)] arranged by a wet-phase inversion method, when as templates the enantiomers of phenylalanine were used (separately) [62]. They reported that for D-/L-Phe-imprinted membranes and after 16 mL (34 min) of ultrafiltration: the adsorption selectivity of 2.25 / 2.40, the rejection selectivity of 0.13 / 0.28, and the permselectivity of 1.94 / 2.08, were achieved. According to them, the membrane was selective.

CONCLUSION

Phenylalanine is an α-amino acid containing the aromatic ring in molecule and it is used as a precursor in tyrosine synthesis and other compounds with the six-membered aromatic ring, like dopamine, and catecholamines: adrenaline, noradrenaline and melanine. Phenylketonuria is a genetic (autosomal, recessive) inborn error of metabolism causing by lack of phenylalanine hydroxylase (PAH). Non-diagnosed and non-treated PKU can cause hyperphenylalaninemia (HPA) and impaired postnatal cognitive development. In PKU treatment, a diet with low phenylalanine concentration is used. Phenylalanine is metabolized into number of compounds, like tyramine, β-phenylethylamine (β-PEA), tryptamine and octopamine, which are biologically active amines. Additionally, to this group may also be classified amphetamine and its analogues (for instance, methamphetamine, phenylpropanolamine and pseudoephedrine). D-phenylalanine has a secondary biological and physiological role – it may be a potential drug in the PKU therapy, and D-phenylalanine derivatives may be naturally-occurring sweeteners, or D-phenylalanine can be used in new drugs synthesis. For the enantiomers separation process of amino acids, including phenylalanine, mostly enantioseparation is used, and membranes like: chiral thermo-sensitive membranes or channel type, and solution-diffusion (stationary phase in column chromatography may contain chiral polymers, DNA and others).

ACKNOWLEDGMENT

Thank you very much for M.Sc. Michalina Zajdel (The Maria Skłodowska-Curie Institute of Oncology) for text review and helpful comments.

REFERENCES

[1] Zurflüh M. R., Zschocke J., Lindner M., Feillet F., Chery C., Burlina A., Stevens R. C., Thöny B., Blau N. *Hum Mutat.* 2008 Jan; 29 (1):167-75.

[2] Steinfeld R., Kohlschütter A., Ullrich K., Lukacs Z. *J Inherit Metab Dis.* 2004; 27(4): 449-53.

[3] Guthrie R., Susi A. *Pediatrics* 1963 Sep; 32: 338-43.

[4] http://www.ncbi.nlm.nih.gov/gene/?term=PAH
[5] Gibbs R. *Science*, Vol 316, 13 Apr. 2007, 222-234.
[6] Blau N., van Spronsen F. J., Levy H. L. *Lancet* 2010 Oct 23; 376 (9750):1417-27.
[7] Kaufman S., Max E. E., Kang E. S. *Pediatr Res.* 1975 Aug; 9 (8): 632-4.
[8] Kalaydjieva L., Dworniczak B., Aulehla-Scholz C., Devoto M., Romeo G., Sturhmann M., Kucinskas V., Yurgelyavicius V., Horst J. *J Med Genet.* 1991 Oct; 28(10): 686-690.
[9] Muntau A. C., Röschinger W., Habich M., Demmelmair H., Hoffmann B., Sommerhoff C. P., Roscher A. A. *N. Engl J. Med.* 2002 Dec 26; 347(26): 2122-32.
[10] Matalon R., Koch R., Michals-Matalon K., Moseley K., Surendran S., Tyring S., Erlandsen H., Gamez A., Stevens R. C., Romstad A., Møller L.B., Guttler F. *Genet Med.* 2004 Jan-Feb; 6(1): 27-32.
[11] Thompson A. J., Tillotson S., Smith I., Kendall B., Moore S. G., Brenton D. P. *Brain* 1993 Aug; 116 (Pt 4): 811-21.
[12] Griffiths P. V., Demellweek C., Fay N., Robinson P. H., Davidson D. C. *Arch Dis Child.* 2000 Mar; 82(3): 209-15.
[13] Weglage J., Grenzebach M., Pietsch M., Feldmann R., Linnenbank R., Denecke J., Koch H. G. *J Inherit Metab Dis.* 2000 Jul; 23(5): 487-96.
[14] Levy H. L., Lobbregt D., Barnes P. D., Poussaint T. Y. *J Pediatr.* 1996 Jun; 128 (6): 770-5.
[15] Rouse B., Azen C., Koch R., Matalon R., Hanley W., de la Cruz F., Trefz F., Friedman E., Shifrin H. *Am J Med Genet.* 1997 Mar 3; 69 (1): 89-95.
[16] Waisbren S. E., Hanley W., Levy H. L., Shifrin H., Allred E., Azen C., Chang P. N., Cipcic-Schmidt S., de la Cruz F., Hall R., Matalon R., Nanson J., Rouse B., Trefz F., Koch R. *JAMA.* 2000 Feb 9; 283(6): 756-62.
[17] Rouse B., Matalon R., Koch R., Azen C., Levy H., Hanley W., Trefz F., de la Cruz F. *J Pediatr.* 2000 Jan; 136(1): 57-61.
[18] Levy H. L., Guldberg P., Güttler F., Hanley W. B., Matalon R., Rouse B. M., Trefz F., Azen C., Allred E. N., de la Cruz F., Koch R. *Pediatr Res.* 2001 May; 49(5): 636-42.
[19] Brumm V. L., Bilder D., Waisbren S. E. *Mol. Genet. Metab.* 2010; 99 Suppl 1: S59-63.
[20] Scriver C. R., Gregory D. M., Sovetts D., Tissenbaum G. *Metabolism* 1985 Sep; 34(9): 868-73.

[21] van Spronsen F. J., Reijngoud D. J., Smit G. P., Nagel G. T., Stellaard F., Berger R., Heymans H. S. *J Clin Invest.* 1998, Jun 15; 101(12): 2875-80.

[22] Thompson G. N., Halliday D. *J. Clin. Invest.* 1990, 86: 317-322.

[23] Thompson G. N., Walter J.H., Leonard J.V., Halliday D. *Metabolism* 1990, 39: 799-807.

[24] Treacy E., Pitt J. J., Seller K., Thompson G. N., Ramus S., Cotton R. G. H. 1996. *J. Inherited Metab. Dis.* 19: 595-602.

[25] Hanley W. B., Lee A. W., Hanley A. J., Lehotay D. C., Austin V. J., Schoonheyt W. E., Platt B. A., Clarke J. T. *Mol Genet Metab.* 2000 Apr; 69(4): 286-94.

[26] Leuzzi V., Bianchi M. C., Tosetti M., Carducci C. L., Carducci C. A., Antonozzi I. *J. Inherit Metab Dis.* 2000 Sep; 23(6): 563-70.

[27] Weglage J., Pietsch M., Denecke J., Sprinz A., Feldmann R., Grenzebach M., Ullrich K. *J Inherit Metab Dis.* 1999 Aug; 22(6): 693-705.

[28] Koch R., Burton B., Hoganson G., Peterson R., Rhead W., Rouse B., Scott R., Wolff J., Stern A. M., Guttler F., Nelson M., de la Cruz F., Coldwell J., Erbe R., Geraghty M. T., Shear C., Thomas J., Azen C. *J Inherit Metab Dis.* 2002 Sep; 25(5): 333-46.

[29] Stegink L. D., Filer L. J. Jr, Baker G. L., Bell E. F., Ziegler E. E., Brummel M. C., Krause W. L. *Metabolism* 1989 Jan; 38(1): 78-84.

[30] Glushakov A. V., Dennis D. M., Morey T. E., Sumners C., Cucchiara R. F., Seubert C. N., Martynyuk A. E. *Molecular Psychiatry* 7 (2002) 359-367.

[31] Glushakov A. V., Glushakova O., Varshney M., Bajpai L. K., Sumners C., Laipis P. J., Embury J. E., Baker S. P., Otero D. H., Dennis D. M., Seubert C. N.. Martynyuk A. E. *Brain* 128 (2005) 300-307.

[32] Broadley K. J. *Pharmacology & Therapeutics* 125 (2010) 363-375.

[33] Burchett S. A., Hicks, T. P. *Prog Neurobiol.* 79 (2006) 223-246.

[34] Branchek T., Blackburn, T. P. *Curr Opin Pharmacol* 3 (2003), 90-97.

[35] Borowsky B., Adham N., Jones K. A., Raddatz R., Artymyshyn, R., Ogozalek K. L. *Proc Natl Acad Sci USA* 98 (2001) 8966-8971.

[36] Zucchi R., Chiellini G., Scanlan, T. S., & Grandy, D. K. *J Pharmacol* 149 (2006) 967-978.

[37] Burchett S. A., & Hicks T. P. *Prog Neurobiol* 79 (2006) 223-246.

[38] Bunzow J. R., Sonders M. S., Arttamangkul S., Harrison L.M., Zhang G., Quigley D. I. *Mol Pharmacol* 60 (2001) 1181-1188.

[39] Branchek T., & Blackburn T. P. *Curr Opin Pharmacol* 3 (2003) 90-97.

[40] Lindemann L., Hoener M. C. *Trends Pharmacol Sci* 26 (2005) 274-281.
[41] Sulzer D. et al. / *Progress in Neurobiology* 75 (2005) 406-433.
[42] Burchett S. A., Hicks T. P. / *Progress in Neurobiology* 79 (2006) 223-246.
[43] Lindemann L., Hoener M. C. *TRENDS in Pharmacological Sciences* Vol. 26 No. 5 May 2005.
[44] Frascarelli S., Ghelardoni S., Chiellini G., Vargiu R., Ronca-Testoni S., Scanlan T. S., Grandy D. K., Zucchi R. *European Journal of Pharmacology* 587 (2008) 231-236.
[45] Singh V., Rai R. K., Arora A., Sinha N., Thakur A. K. *Nature / Scientific Reports* | 4 : 3875 | DOI: 10.1038/srep03875.
[46] Adler-Abramovich L. et al. *Nat. Chem. Biol.* 8, 701-706 (2012).
[47] Olsztyńska S., Komorowska M. *Struct. Chem.* 23, 1399-1407 (2012).
[48] Christianson D. W., Mangani S., Shoham G., Lipscomb W. N. *The Journal of Biological Chemistry* Vol. 264, No. 22, Issue of August 5, pp. 12849-12853, 1989.
[49] Dhaneshwar S. S., Chail M., Patil M., Naqvi S., Vadnerkar G. *European Journal of Medicinal Chemistry* 44 (2009) 131-142.
[50] Masuda K., Koizumi A., Misaka T., Hatanaka Y., Abe K., Tanaka T., Ishiguro M., Hashimoto M. *Bioorganic & Medicinal Chemistry Letters* 20 (2010) 1081-1083.
[51] Zhang L., Xu J., Huang Y., Min S. *Spectrochimica Acta Part A* 74 (2009) 835-838.
[52] Dhaneshwar S., Buchade R., Sutari A., Karandikar H., Tewari K. *International Journal of Pharma and Bio Sciences* Vol 2/Issue 2/Apr-Jun 2011, p. 330-337.
[53] Lovelock S. L., Lloyd R. C., Turner N. J. *Angew. Chem. Int. Ed.* 2014, 53, 4652-4656.
[54] Solanas C., De La Torre B. G., Fernández-Reyes M., Santiveri C. M., Jiménez M. Á., Rivas L., Jiménez A. I., Andreu, D., Cativiela C. *Journal of Medicinal Chemistry* 52 (3), 12 February 2009, p. 664-674.
[55] Zhang S., Ding J., Liu Y., Kong J., Hofstetter O. *Analytical Chemistry* 78 (21), 1 November 2006, p. 7592-7596.
[56] Khan H., Khan T., Park J. K. *Separation and Purification Technology* 62 (2008) 363-369.
[57] Feng X., Zhang Q., Liang X., Li J., Zhao Y., Chen L. *J. Polym. Res.* (2014) 21: 475.
[58] Matsuoka Y., Kanda N., Lee Y. M., Higuchi A., *Journal of Membrane Science* 280 (2006) 116-123.

[59] Higuchi A., Hayashi A., Kanda N., Sanui K., Kitamura H., *J. Mol. Struct.* 739 (2005) 145-152.

[60] Higuchi A., Higuchi Y., Furuta K., Yoon B. O., Hara M., Maniwa S., Saitoh M., Sanui K. *J. Membr. Sci.* 221 (2003) 207-218.

[61] Trost E.-M., Fischer L. *Journal of Molecular Catalysis B: Enzymatic* 19-20 (2002) 189-195.

[62] Ul-Haq N., Khan T., Park J. K. 83 (4), *Journal of Chemical Technology and Biotechnology* April 2008, Pages 524-533.

In: Phenylalanine
Editor: Darlene Williamson

ISBN: 978-1-63482-501-6

Chapter 2

PHENYLKETONURIA'S IMPACT ON PHYSICAL GROWTH IN A SPANISH COHORT

Luis Aldámiz-Echevarría[1,*], Marta Llarena[1], Fernando Andrade[1], María L. Couce[2], Domingo González-Lamuño[3], María A. Bueno[4], María C. García[5], Jaime Dalmau[6], Isidro Vitoria[6], Javier Blasco[7], Carlos Alcalde[8], David Gil[9], Mónica Ruiz[10], Luis Peña-Quintana[11], María A. Ruiz[12], David González[1] and Felix Sánchez-Valverde[14]

[1]Metabolism Unit, Cruces University Hospital, Barakaldo, Vizcaya, Spain

[2]Metabolic Disorders Unit, Santiago de Compostela University Hospital, Santiago de Compostela, La Coruña, Spain

[3]Nephrology and Metabolism Unit, Marqués de Valdecilla University Hospital, Santander, Spain

[4]Metabolic Disorders, Dietetics and Nutrition Unit, Virgen del Rocío University Hospital, Sevilla, Spain

[5]Metabolic Pathologies Unit, Miguel Servet University Hospital, Zaragoza, Spain

[6]Nutrition and Metabolopathologies Unit, La Fe University Hospital, Valencia, Spain

[*] Corresponding autor: E-mail: luisjose.aldamiz-echevarazuara@osakidetza.net

[7]Gastroenterology, Hepatology and Child Nutrition Unit, Carlos Haya University Hospital, Málaga, Spain
[8]Paediatrics Unit, Río Hortega University Hospital,Valladolid, Spain
[9]Gastroenterology Unit, Virgen de la Arrixaca University Hospital, Murcia, Spain
[10]Paediatrics Unit, Nuestra Señora de la Candelaria University Hospital, Santa Cruz de Tenerife, Spain
[11]Paediatric Gastroenterology, Hepatology and Nutrition Unit, Mother and Child Hospital Complex, Las Palmas de Gran Canaria, Spain
[12]Metabolic Pathologies and Neuropaediatrics Unit, Son Espases University Hospital, Palma de Mallorca, Spain
[13]Metabolic Pathologies Unit, Maternal and Child Hospital, Badajoz, Spain
[14]Gastroenterology and Paediatric Nutrition Unit, Virgen del Camino Hospital, Pamplona, Spain

ABSTRACT

Phenylketonuria is an inborn error of metabolism affecting the phenylalanine metabolic pathway, which converts phenylalanine to tyrosine via phenylalanine 4-hydroxylase and its cofactor (6R)-L-erythro-5,6,7,8-tetrahydrobiopterin (BH_4). Phenylketonuria treatment is based on strict vegetarian diets, with very low phenylalanine intake and supplemented with phenylalanine-free formulas. This phenylalanine-restricted diet has proven to be effective in preventing the development of long-term neurological damage caused by phenylalanine accumulation. However, such diets have occasionally been reported to hinder normal development, since some individuals presented growth retardation. Furthermore, in an attempt to improve outcomes in these patients, several alternative approaches for the treatment of phenylketonuria have emerged, such as therapy based on administration of the cofactor BH_4. This treatment allows patients to consume a near normal diet, or at least, a less strict low-protein diet, which results in higher natural protein intake. However, little is known about how BH_4 treatment affects physical development.

Firstly, in order to evaluate the impact of the phenylketonuria diet on anthropometric characteristics (weight, height, body mass index and growth rate), we conducted a retrospective longitudinal study. The aim was to further explore the hypothesis of higher natural protein intake being associated with attaining improved physical outcomes.

Anthropometric characteristics and nutrition were evaluated from birth to adulthood in a cohort of phenylketonuria and mild-hyperphenylalaninaemia patients, who were exclusively on protein-restricted diets, and were compared to the Spanish reference population. Patients with phenylketonuria showed growth impairment in early stages, with higher phenylalanine intakes being associated with improved developmental outcomes over this period. Our results suggest that prescribing very stringent diets in early stages might predispose these patients to later growth retardation, with growth outcomes in adulthood being well below the 50th percentile for healthy subjects.

Secondly, we conducted a retrospective longitudinal study of anthropometric characteristics in a cohort of patients with phenylketonuria on BH_4 treatment and compared their developmental outcomes with those of a group of patients on a phenylalanine-restricted diet, in order to determine whether BH_4 treatment was associated with an improvement in growth development. Data were collected every 6 months over two periods of time (two and five years). No improvement was observed in the anthropometric variables in the BH_4-treated group, from prior to initiating treatment to when they had taken the drug for 2 or 5 years. In addition, growth impairment was also observed in patients on low-phenylalanine diets. In fact, individuals on long-term BH_4 treatment seemed to achieve similar developmental outcomes to those on more restricted diets. In conclusion, our study identified growth impairment in patients with phenylketonuria on BH_4 treatment, despite the fact that their natural protein intake increased.

INTRODUCTION

Phenylketonuria (PKU; OMIM 261600) is an autosomal recessive inborn error of metabolism affecting the phenylalanine (Phe) metabolic pathway, which converts Phe to tyrosine via phenylalanine 4-hydroxylase (PAH, EC 1.14.16.1) and its cofactor (6R)-L-erythro-5,6,7,8-tetrahydrobiopterin (BH_4). According to the residual activity of the mutant enzyme, patients are classified into four different phenotypes: mild-hyperphenylalaninaemia (HPA), which does not require further treatment; mild-PKU; moderate-PKU and classic-PKU. If untreated, PKU can result in impaired cognitive development as a consequence of the neurotoxic effect of Phe accumulation [1].

Traditionally, PKU treatment consists of strict vegetarian diets, with very low Phe intake and supplemented with Phe-free protein substitutes and specially-manufactured low-protein foods. This Phe-restricted diet aims to maintain blood Phe levels below pre-established "safe" thresholds; and has proven to be effective in preventing the development of long-term neurological damage caused by Phe accumulation [2, 3]. However, such diets have occasionally been reported to hinder normal development, since some individuals presented growth retardation and malnutrition [4-6]. Several studies have attempted to establish a connection between a reduction in natural protein intake and growth retardation, but their findings have been inconsistent. Therefore, we firstly conducted a retrospective longitudinal study, in order to evaluate the impact of PKU protein-restricted diet on anthropometric characteristics (weight, height, body mass index and growth rate). The aim was to further explore the hypothesis of higher natural protein intake being associated with attaining improved physical outcomes in patients with PKU. In this study, anthropometric characteristics and nutrition were evaluated from birth to adulthood in a cohort of patients with PKU and mild-HPA, who were exclusively on protein-restricted diets; and were compared to the Spanish reference population [7]. The results and conclusions of this study are shown in further detail in the Results' section (Study 1).

Treating patients with PKU represents a huge nutritional challenge. In the last decades, several alternative approaches for the treatment of PKU have emerged in an attempt to improve outcomes in these patients; for example, large neutral amino acid supplementation, BH_4 therapy, novel protein sources (such as glycomacropeptide) and enzyme substitution therapy using phenylalanine ammonia lyase [8]. These strategies focuses on maintaining blood Phe levels within pre-established "safe" thresholds, ameliorating secondary complications as a result of daily nutrient intake below recommended dietary allowances (RDAs) [9], as well as increasing patients' dietary adherence. PAH-deficient patients positively responding to BH_4 treatment were observed for the first time in 1999 by Kure et al. [10]. Therapy based on this cofactor has opened up new treatment possibilities for a significant proportion of these patients, namely those responding to a BH_4 loading test. So far, clinical data on BH_4 treatment's benefits have highlighted a remarkable increase in Phe tolerance, a reduction in the fluctuation of blood Phe levels and an improvement in nutritional status and quality of life [9, 11-13]. In addition, BH_4 treatment allows patients with PKU to partially or totally liberalise their Phe-restricted diet, reducing their need to be supplemented with protein substitutes and enabling them to increase their intake of natural proteins. However, so far there is little clinical evidence regarding the impact of BH_4 treatment on PKU patient physical growth and the results from the few published studies have been inconclusive [14, 15]. Therefore, we secondly conducted a retrospective longitudinal study of anthropometric characteristics in a cohort of patients with PKU on BH_4 treatment and compared their developmental outcomes with those of a group of patients on a Phe-restricted diet, in order to determine whether BH_4 treatment was associated with an improvement in growth development. We also assessed nutritional status, in order to further investigate the association of higher natural protein intake with improved developmental outcomes [16]. The results and conclusions of this study are shown in further detail in the Results' section (Study 2).

RESULTS

Study 1: Anthropometric characteristics and nutrition in a Spanish cohort of PAH-deficit patients exclusively treated with protein-restricted diets.

We conducted a retrospective, longitudinal, multicentre study of PAH-deficient patients, who were regularly treated and monitored at several Spanish hospitals. Patients were classified into one of the four phenotypic categories according to their blood Phe levels, which were measured at diagnosis: classic PKU (Phe > 1200 μmol/L), moderate PKU (Phe: 600-1200 μmol/L), mild PKU (Phe: 360-600 μmol/L) and mild-HPA (Phe < 360 μmol/L). The inclusion criteria for the study were: (1) early diagnosis of PKU or mild-HPA; (2) in the case of patients on a PKU diet, exclusive and continuous treatment with Phe-restricted diet, supplemented with Phe-free substitutes and specially manufactured low-protein foods, the diet being initiated within days or weeks of birth; (3) absence of any other diseases known to affect physical development; (4) Caucasian race; and (5) regular attendance to their scheduled clinical check-ups. A total of 505 PAH-deficient patients were included in the present study: 279 were diagnosed with PKU (130 males, 149 females), and 226 with mild-HPA (106 males, 120 females). The PKU group consisted of 43 patients with mild PKU, 78 with moderate PKU and 158 with classic PKU. Their ages ranged from 1 to 36 years.

Clinical, biochemical and nutritional data were collected for the first 18 years of life in patients with PKU and for the first 12 years of life in patients with mild-HPA. The anthropometric characteristics which were assessed in our study were: height, weight, body mass index (BMI) and growth rate (GR), which were converted into Z-scores by subtracting the expected mean measurement for the corresponding age and gender and dividing by its standard deviation. BMI was calculated using the formula BMI = weight (kg)/height^2 (m^2). GR was calculated from two height measurements, which were taken approximately 1 year apart. We used the 2008 Spanish Growth Studies as the source of data for the reference population [17, 18]. The statistical analysis (IBM SPSS® 21.0) indicated that mean Z-score for weight gradually decreased during the first 2 years of life to a nadir of -0.59 ± 0.95 and -0.51 ± 0.99 in male and female patients with PKU, respectively. From 2 to 3.5 years of age, it gradually increased and remained almost steady until 16.5 years of age (mean Z-scores for weight close to 0 and -0.2, in the male and female group, respectively). In adulthood, mean Z-score for weight in the male group fell well below the 50th percentile for healthy subjects (-0.51 ± 0.82), and it rose above this threshold in the female group (0.34 ± 1.13). In contrast, mean Z-score for weight in patients with mild-HPA remained close to 0 over the observation period. With regard to height Z-scores in the PKU group, they behaved similarly to weight Z-scores. They fell well below 0 during the first 2 years of life (mean Z-scores for height: -0.87 ± 1.14 and -0.68 ± 1.16 in the males and female group, respectively), and then recovered slightly towards 0. From 16.5 to 18 years, mean Z-score for height in the male group dropped to a final value of -0.89 ± 1.03. In the female group, mean Z-scores for height decreased steadily from 12 years of age to adulthood (final value -0.88 ± 0.81). On the other hand, mean Z-scores for height in male patients with mild-HPA were above the 50th percentile for the healthy population, but fell to -0.49 ± 1.06 at 12 years of age; while in the female group, it fell to a nadir of -0.59 ± 1.27 at 7.5 years of age, and was close to 0 at 12 years of age. Our results are consistent with the findings of several previous European studies on physical development in patients with PKU [19-21]. In marked contrast to our findings, a Spanish study has observed normal developmental outcomes in PAH-deficient patients over an observation period from 0 to 18 years of age; even in the most severe phenotypes, who usually are on a strictest diet [22].

No significant variation was observed in the evolution of BMI Z-scores in the male PKU group, its mean value was close to the 50th percentile for healthy subjects over the observation period. However, the female PKU group showed an increase in the mean Z-score for BMI at 8 years of age (0.98 ± 0.93); and, in contrast to the male group, BMI Z-score was well above 0 at 18 years of age. Mean Z-scores for BMI in the mild-HPA group were approximately 0 from 6 months to 12 years of age, in both the male and female groups. Finally, mean Z-scores for GR in the PKU group showed a gradual decrease between 1.5 and 4.5 years of age, and a sharp fall around puberty in both sub-groups. In the mild-HPA group, mean Z-scores for GR fell steadily in early stages of growth, and remained above the 50th percentile for healthy subjects until 11 years of age, dropping below this threshold from this age onwards.

We also performed paired-sample *t*-tests, in order to analyse intra-group (male vs. female) and inter-group (male/female-PKU vs. male/female-mild-HPA) differences in anthropometric characteristics. Statistically significant differences in mean Z-scores for height and GR (PKU group: male vs. female) were observed using the Student's *t*-test ($p < 0.05$). In the mild-HPA group, we found significant differences in BMI Z-scores between male and female patients ($p < 0.05$).

Blood Phe concentrations were measured at diagnosis and every 6 months thereafter, in order to evaluate the effectiveness of the prescribed diet for each patient and monitor their adherence to the treatment. According to the Spanish protocol for treating and monitoring patients with PKU [23, 24], blood Phe levels should remain below cut-offs established for each age group: < 360 μmol/L for individuals < 6 years of age; < 480 μmol/L for those from 6 to ≤ 10 years of age; and < 600 μmol/L for those > 10 years of age. As we expected, mean blood Phe levels were significantly higher in the PKU group than in the mild-HPA group. However, no significant differences were observed in blood Phe levels between the male and female groups.

With regard to nutrition, dietary intervention was started at diagnosis. Patients were on Phe-restricted diets, supplemented with Phe-free amino acid mixtures and specially-manufactured low-protein foods. Specifically, they had a restricted intake of protein-rich foods (such as meat, fish, eggs, milk and dairy products, and legumes) as well as foods with moderate protein content (such as cereals). The daily amount of natural protein prescribed for each individual was modified according to their age and the most recent blood Phe values. The recommended daily intake of protein substitutes was also based on patients' age and they were advised to divide it into at least four daily doses. Dietary parameters (such as intakes of natural protein, protein contained in amino acid mixtures, total protein and Phe) were calculated by an experienced dietician from the 3-day food records completed by parents/legal guardians. In the present study, nutritional data were analysed every 6 months from birth to 18 years of age for 98 out of the 279 patients with PKU; but not in the case of patients with mild-HPA due to incomplete records.

The intake of the abovementioned nutrients was analysed by age (0-2; 2-9; 9-12 and 12-18 years of age) (Table 1). The statistical analysis indicated that median intakes of natural protein, protein contained in amino acid mixtures

Table 1. Dietary parameters (intakes of natural protein, protein contained in amino acid mixtures, total protein and Phe) in the PKU group. Values are expressed as median (interquartile range)

Age (years)	Natural protein (g/kg/day)	Protein in amino acid mixtures (g/kg/day)	Total protein (g/kg/day)	Phe (mg/kg/day)
0-2	0.8 [0.5-1.0]	1.4 [1.0-1.7]	2.2 [1.8-2.8]	28.2 [20.8-36.2]
2-9	0.5 [0.3-0.9]	1.3 [1.0-1.7]	1.8 [1.4-2.3]	15.5 [10.9-20.4]
9-12	0.3 [0.2-0.5]	1.2 [0.9-1.4]	1.2 [1.0-1.7]	8.6 [7.0-11.3]
12-18	0.3 [0.2-0.5]	1.0 [0.7-1.3]	1.1 [1.0-1.4]	7.6 [5.4-10.3]

Phe: Phenylalanine; PKU: Phenylketonuria.

Total protein: combination of both natural protein and protein contained in amino acid mixtures.

and total protein met the recommendations in the Spanish protocol for treating PKU in each group of age [23, 24]. However, natural protein intake was below the recommended average value of the World Health Organisation (WHO) for healthy individuals (0-18 years of age); whereas the intakes of the other dietary components containing protein were above the guidelines for normal growth [25]. With regard to Phe intake, its median met the recommendations in the Spanish protocol for treating PKU in each group, but was lower than the guidelines of the WHO. In addition, we further evaluated the influence of the Phe-restricted diet on physical growth in patients with PKU using bivariate correlation analysis. There was no association between higher protein intakes and attaining improved developmental outcomes in the PKU group. On the other hand, we observed a significant positive correlation between higher Phe intakes and achieving better height Z-scores in the PKU female group (Pearson's correlation coefficients (r)= 0.76, 0.56 and 0.69, $p < 0.05$, at 1.5, 2.5 and 4.5 years of age, respectively) as well as in the PKU male group (r= 0.61, 0.73, 0.78, 0.74, 0.86, 0.72 and 0.83, $p < 0.05$, at 2.5, 3.5, 4.0, 4.5, 5.0, 5.5 and 11 years of age, respectively).

In addition, weight Z-scores positively correlated with Phe intake in the PKU female group (r= 0.52 and 0.46, $p < 0.05$, at 1.5 and 2.5 years of age, respectively) and in the PKU male group (r= 0.68, 0.61, 0.74 and 0.78, $p < 0.05$, at 3.5, 4.0, 4.5 and 5.0 years of age, respectively).

In summary, our findings from study 1 showed growth impairment in the group of patients with PKU. In particular, there were two well-differentiated periods of age (from birth to 2 years of age, and from end of growth to adulthood) when height Z-score fell well below 0.

We also found a significant positive correlation between height and Phe intake, as well as between weight and Phe intake. On the contrary, in patients with mild-HPA, no growth retardation was observed.

Study 2: Impact of phenylketonuria on growth in BH_4-treated and Phe-restricted diet-only patients.

Patients with PKU below 17 years of age were included in this retrospective longitudinal study from thirteen hospitals in Spain. The study included patients with classic, moderate and mild PKU. They were divided into two groups according to their treatment: (1) BH_4 treatment (until 2009 as [6R]-L-erythro-5,6,7,8-tetrahydrobiopterin [Schircks Laboratories, Jona, Switzerland], and thereafter sapropterin dihydrochloride [KUVAN®, Merck, Madrid, Spain]), and (2) exclusively Phe-restricted diet. The inclusion criteria for this study were: (1) early diagnosis, confirmed by mutation analysis of

PAH gene; (2) in the case of patients on a PKU diet, early and continuous treatment with Phe-restricted diet, supplemented with Phe-free substitutes and specially manufactured low-protein foods; (3) in the case of BH_4 patients, responsiveness to BH_4 treatment (a defect in the BH_4 pathways being ruled out by analysing urinary pterin levels); (4) absence of any other diseases known to affect physical development; (5) Caucasian race and (6) regular attendance to their scheduled clinical check-ups.

Generally, BH_4 responsiveness is evaluated using a BH_4 loading test. The BH_4 loading test used in this study was based on recommendations in the Spanish protocol for treating and monitoring patients with PKU [7, 16, 23, 24]. Briefly, participants were loaded with Phe before BH_4 administration. In general, an individual was considered to be a primary responder to BH_4 treatment when blood Phe levels fell by at least 30% within 8 h after BH_4 administration. Patients who showed a reduction in blood Phe levels of at least 30% within 12 h after BH_4 administration were classified as late responders. From 2005 onwards, BH_4 responsiveness was evaluated based upon the 50% criterion at Virgen del Rocío University Hospital (for a full description of the protocol, see reference [26]). This BH_4 loading test was designed as a two-stage protocol. In the first stage (the 24-h test), participants were loaded with Phe before BH_4 administration. With this initial stage of the protocol, an individual was considered BH_4 responsive when blood Phe levels fell by at least 50% within 24 h after BH_4 administration. PKU patients who did not meet the aforementioned criterion for responsiveness underwent the therapeutic test. In this second stage of the protocol, patients were administered a BH_4 dose of 20 mg/kg per day for one week together with a daily intake of protein, set on a case-by-case basis in line with age- and gender-specific RDAs. The results of the therapeutic test were considered to be positive when Phe levels remained below an established threshold (< 360 μmol/L for individuals < 6 years of age; < 480 μmol/L for those from 6 to $\leq$ 10 years of age; and < 600 μmol/L for those > 10 years of age). Patients who met this criterion were classed as late responders.

In this second study, weight and height were collected over a period of two or five years, every 6 months, from the date when BH_4 treatment was initiated. The same timetable was used in the diet-only group, in order to compare the evolution of anthropometrics and nutrition between both populations. Height, weight, BMI and GR measurements were converted into Z-scores by subtracting the expected mean measurement for the corresponding age and gender and dividing by its standard deviation. We used the 2008

Spanish Growth Studies as the source of data for the reference population [17, 18].

Blood Phe concentrations were regularly measured as described in study 1. Samples for Phe monitoring were scheduled to be collected according to patients' age: weekly from 0 to 6 months of age, every other week from 6 months to 2 years of age, and monthly from 2 years of age onwards. Blood Phe levels should remain below cut-offs established for each age group, according to the Spanish protocol for treating and monitoring patients with PKU [23, 24], as indicated in the previous section.

Diet-only patients were on Phe-restricted diets, supplemented with Phe-free amino acid mixtures and specially manufactured low-protein foods (further details are described in the sub-section Study 1). With regard to BH_4 patients, in an attempt to enable them to consume a normal diet, they gradually increased their intake of certain protein-rich foods, while maintaining BH_4 treatment, and consequently they reduced their intake of protein in amino acid mixtures. Although patients with a Phe reduction below 50% in the BH_4 loading test could not adopt a normal diet, they obtained at least 70% of total protein from natural sources. Dietary parameters (such as intakes of natural protein, protein contained in amino acid mixtures, total protein and Phe) for both groups were calculated as described in the study 1.

The statistical analysis (IBM SPSS® 21.0) was carried out considering two periods: two-year and five-year follow-up.

Two-Year Follow-Up

Follow-up data from patients with PKU were collected every 6 months over a period of 2 years. The group on BH_4 treatment included 36 patients (18 males and 18 females): 5 with classic PKU, 24 with moderate PKU and 7 with mild PKU. This BH_4 group represented the 33% of total patients with PKU included in this study. The rest of the cohort (72 patients: 36 males and 36 females) was on exclusively Phe-restricted diet supplemented with amino acids. There were 10 patients with classic PKU, 48 with moderate PKU and 14 with mild PKU. Thus, there were two diet-only patients for each BH_4 patient matched for age and type of PKU. Overall, there were 54 females and 54 males with an initial age range from 6 months to 15 years.

The comparison of anthropometric characteristics between the BH_4 and the diet-only group is shown in Table 2. The negative values of weight and height Z-scores in both groups, which were below the Spanish healthy population, may reflect the impact of a Phe-restricted diet until BH_4 treatment was initiated in one of the groups.

Table 2. Anthropometric characteristics (weight, height, BMI and GR Z-scores) in the BH_4-treated group comparing with the diet-only group over the two-year follow-up period

	Weight Z-score		Height Z-score		BMI Z-score		GR Z-score	
	BH_4	diet-only	BH_4	diet-only	BH_4	diet-only	BH_4	diet-only
Initial	-0.16±1.03	-0.55±0.83	-0.71±1.25	-0.76±1.03	0.31±1.03*	-0.17±0.88*	0.15±2.18	0.02±2.37
6 months	-0.08±0.99*	-0.56±0.89*	-0.62±1.13	-0.74±0.99	0.28±0.82*	-0.19±0.96*	0.61±2.15	-0.01±1.23
12 months	-0.13±1.13*	-0.71±0.7*	-0.69±1.1	-0.91±0.88	0.33±1.21*	-0.27±0.77*	0.08±1.67	0.35±1.59
18 months	-0.79±1.17	-0.46±0.9	-0.72±1.12	-0.81±0.98	0.51±1.22*	-0.14±0.87*	-0.01±1.52	0.09±1.73
Final	-0.75±1.04*	-0.52±0.85*	-0.73±0.96	-0.90±1.00	0.37±1.09*	-0.12±0.89*	0.05±2.71	0.10±2.26

Statistically significant differences between the groups: BH_4 vs. diet-only group: * ($p < 0.05$).

Initial: Data were collected immediately before initiation of BH_4 treatment. Data were collected every 6 months over the two-year observation period.

Table 3. Dietary parameters (intakes of natural protein, total protein, protein contained in amino acid mixtures and Phe) in PKU patients on BH_4 treatment and on a Phe-restricted diet. Data from the two-year and five-year follow-up studies are included. Values are expressed as median (interquartile range)

	two-year follow-up		five-year follow-up	
	BH_4 group	Diet-only group	BH_4 group	Diet-only group
Phe intake (initial) (mg/kg/day)	29.9 [18.3-52.3]	19.2 [9.3-31.6]	30.8 [24.6-54.8]	9.9 [8.2-21.8]
Phe intake (final) (mg/kg/day)	41.2 [22.9-48.9]	12.9 [7.9-20.3]	38.1 [17.6-47.9]	10.9 [8.8-13.7]
Natural protein intake (initial) (g/kg/day)	0.7 [0.4-1.1]	0.3 [0.3-0.6]	0.8 [0.3-1.6]	0.3 [0.2-0.6]
Natural protein intake (final) (g/kg/day)	0.8 [0.5-1.0]	0.4 [0.3-0.5]	0.9 [0.7-1.1]	0.3 [0.2-0.4]
Protein in amino acid mixtures (initial) (g/kg/day)	0.9 [0.7-1.2]	1.3 [1.0-1.8]	0.7 [0.5-1.2]	1.5 [1.0-1.7]
Protein in amino acid mixtures (final) (g/kg/day)	0.5 [0.2-1.3]	1.2 [0.9-1.8]	0.3 [0.0-0.7]	1.1 [0.8-1.3]
Total protein (initial) (g/kg/day)	1.8 [1.0-3.6]	2.0 [1.3-2.4]	1.7 [1.3-2.2]	2.0 [1.6-2.4]
Total protein (final) (g/kg/day)	1.5 [0.7-2.2]	1.4 [1.0-2.4]	1.2 [0.7-1.7]	1.6 [1.2-1.9]

Table 4. Anthropometric characteristics (weight, height, BMI and GR Z-scores) in the BH_4-treated group comparing with the diet-only group over the five-year follow-up period

	Weight Z-score		Height Z-score		BMI Z-score		GR Z-score	
	BH_4	diet-only	BH_4	diet-only	BH_4	diet-only	BH_4	diet-only
Initial	-0.08±1.08	-0.01±1.12	-0.29±1.31	-0.27±1.20	0.23±1.10	0.42±1.02	1.62±1.62	0.78±1.78
6 months	0.21±1.13	-0.05±0.99	-0.42±1.17	-0.49±1.26	0.07±0.64	0.17±0.79	-1.31±2.06	0.08±1.31
12 months	0.2±1.52	0.07±1.2	-0.29±1.31	-0.44±1.14	0.40±1.49	0.37±1.10	-0.9±1.75	0.03±1.16
18 months	0.17±1.45	-0.27±0.98	-0.26±1.13	-0.57±1.06	0.67±1.25	-0.04±0.78	0.60±0.90	-0.00±1.31
24 months	0.01±1.26	-0.31±0.82	-0.41±1.26	-0.69±0.95	0.20±1.13	-0.01±0.74	-0.65±1.47	-0.60±0.91
30 months	-0.01±1.02	-0.26±0.92	-0.36±0.94	-0.63±1.07	0.18±0.92	0.02±0.73	-1.23±1.07	-0.14±2.17
36 months	-0.18±1.03	-0.24±0.93	-0.70±1.23	-0.68±1.01	-0.21±0.53	-0.08±0.82	-0.50±1.57	0.12±0.94
42 months	-0.01±1.15	-0.19±0.95	-0.48±1.06	-0.59±1.09	0.29±1.03	0.09±0.80	-0.50±2.41	-0.44±1.53
48 months	-0.49±0.94	-0.31±0.87	-0.86±1.01	-0.82±0.87	-0.09±0.72	0.11±1.01	-0.45±1.84	-0.46±1.20
54 months	0.05±1.29	-0.23±0.92	-0.58±1.06	-0.62±1.12	0.31±1.19	0.02±0.85	0.31±2.27	-0.09±1.47
Final	-0.10±1.01	-0.18±0.88	-0.52±0.90	-0.56±1.09	0.12±1.03	0.04±0.80	0.41±1.67	0.34±2.10

No statistically significant differences between groups were observed ($p>0.05$).

Initial: Data were collected immediately before initiation of BH_4 treatment. Data were collected every 6 months over the five-year observation period.

When we compared both groups, there were no significant differences in the mean Z-scores for height over the observation period. With regard to weight, we observed significant differences ($p < 0.05$); while Z-scores were higher in the BH_4 group comparing with the diet-only group at 6 and 12 months; the mean Z-score for weight was higher in the diet-only group than the BH_4 group at 18 and 24 months. BMI Z-scores were significantly higher in the BH_4 group than the diet-only group within the two-year follow-up period ($p < 0.05$). However, this seems not to be related with treatment, since such difference was already significant before treatment was initiated. Consistent with the lack of significant difference in height Z-scores between groups, there was no significant difference in mean Z-scores for GR between them. However, we observed a trend towards decreased GR in the diet-only group comparing with the BH_4 group.

Moreover, we analysed individually each treatment; there was no significant difference ($p > 0.05$) between initial and final values in the mean Z-scores for weight, height, BMI or GR, which indicated that the treatment had no substantial effect on those parameters.

With regard to blood Phe levels, mean values were lower in BH_4-treated patients than in the low-Phe diet group. There were significant differences between both groups at initial levels (255.2 ± 146.8 µmol Phe/L and 418.4 ± 339.6 µmol Phe/L, BH_4 and diet-only group, respectively; p= 0.003) and between initial and final mean values in the BH_4 group (initial: 255.2 ± 146.8 µmol Phe/L and final: 365.5 ± 226.5 µmol Phe/L; p= 0.005). Furthermore, BH_4 administration led to a decrease in the variability of mean blood Phe levels.

It should be highlighted that stabilising blood Phe levels in patients with PKU is crucial to prevent neurocognitive impairment, especially during the first years of life. In this regard, it was described that blood Phe values above pre-established "safe" thresholds and its high variability correlated with neurocognitive impairment and psychosocial dysfunction [2, 3].

With regard to Phe tolerance over the two-year follow-up, we observed that it increased in 28 of the BH_4-treated patients and remained steady in the other 8. Within the group in which Phe tolerance increased, 11 patients gradually managed to adopt a non-restricted diet; while Phe tolerance increased by 329.2 ± 230.2 mg Phe/day in the other 17 patients, who remained on the restricted diet. Daily intake of natural protein slightly increased from prior to initiating BH_4 treatment to when patients had been taken the drug for 2 years (Table 3).

Five-Year Follow-Up

Follow-up data from patients with PKU were collected every 6 months over a period of 5 years. The BH_4 group included 10 patients (9 with moderate PKU and 1 with mild PKU). The rest of the cohort (20 patients: 12 males and 8 females) was on Phe-restricted diet: 18 with moderate PKU and 2 with mild PKU. Thus, there were two diet-only patients for each BH_4 patient, matched for age and type of PKU. The final mean age for the BH_4-treated group was 10.2 ± 3.1 years, and for the diet-only group was 10.2 ± 3.0 years.

The comparison of anthropometric characteristics between the BH_4-treated and the diet-only group is shown in Table 4. As in the two-year follow-up period, mean Z-scores for height in both groups were below the average for healthy population. Although mean Z-scores for weight in the BH_4 group were also higher than in the diet-only group, the difference did not reach statistical significance ($p > 0.05$). Furthermore, there was no statistically significant difference between initial and final mean values within each group for mean Z-scores for weight, height, BMI and GR.

With regard to blood Phe levels, mean values were slightly higher in the diet-only group comparing with the BH_4-treated group, but this difference did not reach statistical significance at any time point ($p > 0.05$). In addition, there was no statistically significant difference between initial and final mean blood Phe levels in either group (BH_4 group, initial: 204.2 ± 143.9 and final: 289.6 ± 30.6 μmol Phe/L; and diet-only group, initial: 391.8 ± 233.5 and final: 440.6 ± 298.4 μmol Phe/L).

In the five-year follow-up study, we observed that Phe tolerance increased in 6 patients, whereas it remained steady in 2 patients. This level of tolerance did not fall below baseline value in any patient. Within the group in which Phe tolerance increased, 2 patients gradually managed to adopt a non-restricted diet; while Phe tolerance increased by 648.7 ± 348.0 mg Phe/day in the other patients, who required a restricted diet. Daily intake of natural protein slightly increased in the BH_4 group at the end of the five-year follow-up, and it was three times higher than in the diet-only group (Table 3).

In summary, our findings from study 2 showed that there was no improvement in the anthropometric variables in the cohort on BH_4 treatment, from prior to initiating the treatment to when they had taken the drug for 2 or 5 years. Hence, in almost all cases there was a fall in the mean Z-score for the anthropometric variables during these periods, although the differences did not reach statistical significance. In addition, growth impairment was noted in patients on low-Phe diets. Furthermore, we found no statistically significant differences between both groups at any time point. It should be noted that individuals on BH_4 treatment increased their natural protein intake and, in some cases, this allowed them to consume normal diets with protein intake meeting the RDAs. However, no association was observed between higher protein intake and growth.

CONCLUSION

Dietary intervention is one of the currently available treatments options for PKU. This approach, which consists of a marked reduction in Phe intake, has shown to prevent severe neurological impairment which was observed in untreated patients. However, while several reports indicated that an excessive reduction in Phe intake resulted in poor developmental outcomes in patients with PKU; others have not found evidence of the PKU diet negatively affecting physical development. Considering the importance of attaining optimal growth in patients with PKU, we conducted this retrospective longitudinal study on the developmental outcomes in a Spanish cohort of PAH-deficient patients (PKU and mild-HPA) who were exclusively treated with protein-restricted diets (Study 1). The main conclusion from this study was that patients with PKU showed growth impairment in early stages; whereas physical growth was not impaired in patients with mild-HPA.

Since BMI remained relatively steady and close to Z-score=0 from birth to 18 years of age; malnutrition can be ruled out as the leading cause of growth impairment in our patients with PKU. Higher protein intakes have been suggested to be related to improved developmental outcomes, although it has not yet been defined which protein fraction (natural protein, protein substitutes or the combination of both [total protein]) exerts the most influence on physical development. In our cohort, we observed that higher Phe intake in the PKU group was associated with attaining improved developmental outcomes in early stages. Therefore, we suggest that prescribing very stringent diets during this period (first 4 years of life) might predispose patients with PKU to later growth retardation, with growth outcomes in adulthood being well below the 50th percentile for healthy subjects. Moreover, it may be possible that insufficient intake of other nutrients, such as sources of energy or trace elements (selenium or zinc), underlie growth impairment in the PKU population, as it has been previously proposed [27]. Unfortunately, no data on these nutrients were available in our study. However, since patients with PKU on glycomacropeptide-diet and patients with mild-HPA seem to have better growth outcomes than patients on a protein-restricted diet, we hypothesize that the most likely cause of growth impairment is the low protein intake. We consider that amino acids in the supplements may be less efficient than natural amino acids, because of faster oxidation and differences in their metabolism. Further research would be required to elucidate the mechanisms that underlie growth impairment in the PKU population, to confirm the association between reduced protein intakes and poorer growth outcomes. In addition, it would be interesting to evaluate epigenetics and creatine deficit and their relationship with growth outcomes in PKU.

Over the last decade, BH_4 therapy has been extensively used to treat PAH-deficient patients who respond to a BH_4 loading test. This treatment has improved the management of these patients [28, 29]. However, very few studies have explored growth outcomes in patients on BH_4 treatment. Among them, Singh et al. reported that height Z-score increased significantly after 2 years on BH_4 treatment [14]. In addition, they observed that natural protein became the main protein source in the diet, although amino acid supplementation was not always discontinued. Hence, our study aimed to assess to what extent growth in patients with PKU was influenced by BH_4 administration, since it allows them to consume larger amounts of natural proteins (Study 2). The main conclusion was that BH_4-treated patients displayed growth impairment, despite the fact that their natural protein intake increased. Moreover, individuals on long-term BH_4 treatment seemed to achieve similar developmental outcomes to those on more restricted diets. In fact, mean Z-score for almost all anthropometric variables in both groups fell to the same extent, reaching similar final values. It should be noted that we also observed growth impairment in patients on protein-restricted diets, which is consistent with previous reports [19-20, 27, 30-31]. However, in contrast to previous studies, increasing natural protein intake did not seem to have a positive effect on developmental outcomes. In fact, both types of diets seemed to lead patients with PKU to physical development impairment, attaining similar growth outcomes at the end of the observation period. As abovementioned, we concluded that prescribing very stringent diets during the first years of life might lead to growth retardation in PKU. Moreover, our findings suggested that BH_4 therapy, along with the subsequent increase in natural protein intake in these patients, might not counteract the negative impact of the deprived diet during the period prior to initiating BH_4 treatment.

One of BH_4 therapy's aims is to enable patients with PKU to consume normal diets and free them from the need to take Phe-free substitutes. It should be mentioned that many patients on BH_4 treatment in our cohort did not manage to liberalise their diet at the end of the observation period. Moreover, no increase in Phe intake was observed in some BH_4 patients in either two-year or five-year follow-ups.

Taking into account these findings, the question arises as to whether the current protocols to assess BH_4 responsiveness, in particular the 30% criterion, lead to the detection of true responders to the cofactor-based treatment. In our opinion, a shift towards a stricter criterion for responsiveness might ensure that responders to BH_4 treatment would not withdraw from this treatment at later stages. In fact, a new protocol for assessing BH_4 responsiveness based on a 50% criterion has been recently published [26].

Analysis of the protocol showed that, with this criterion, all responders to the BH_4 loading test were true responders to long-term BH_4 treatment, and all these patients consumed normal diets with no need to take amino acid supplements.

To the best of our knowledge, this is the first time that anthropometric characteristics have been assessed in such a large cohort of patients with PKU on BH_4 treatment and over longer observation period (five- vs. two-year follow-up) comparing with previously published studies. Nevertheless, we consider that further research would be necessary in order to explore in more detail how BH_4 treatment affects the health status of patients with PKU.

REFERENCES

[1] Zurflüh MR, Zschocke J, Lindner M, Feillet F, Chery C, Burlina A, Stevens RC, Thöny B, Blau N. Molecular genetics of tetrahydrobiopterin-responsive phenylalanine hydroxylase deficiency. *Hum. Mutat.* 2008 Jan;29(1):167-75

[2] Gassió R, Artuch R, Vilaseca MA, Fusté E, Boix C, Sans A, Campistol J. Cognitive functions in classic phenylketonuria and mild hyperphenylalaninaemia: experience in a paediatric population. *Dev. Med. Child. Neurol.* 2005;47:443-448.

[3] Viau KS, Wengreen HJ, Ernst SL, Cantor NL, Furtado LV, Longo N. Correlation of age-specific phenylalanine levels with intellectual outcome in patients with phenylketonuria. *J. Inherit. Metab. Dis.* 2011;34:963-971.

[4] Dokoupil K, Gokmen-Ozel H, Lammardo AM, Motzfeldt K, Robert M, Rocha JC, van Rijn M, Ahring K, Bélanger-Quintana A, MacDonald A. Optimising growth in phenylketonuria: current state of the clinical evidence base. *Clin. Nutr.* 2012 Feb;31(1):16-21.

[5] Demirkol M, Gizewska M, Giovannini M, Walter J. Follow up of phenylketonuria patients. Mol. Genet. Metab. 2011;104 Suppl:S31-9.

[6] MacDonald A, Rocha JC, van Rijn M, Feillet F. Nutrition in phenylketonuria. *Mol. Genet. Metab.* 2011;104 Suppl:S10-8.

[7] Aldámiz-Echevarría L, Bueno MA, Couce ML, Lage S, Dalmau J, Vitoria I, Andrade F, Blasco J, Alcalde C, Gil D, García MC, González-Lamuño D, Ruiz M, Peña-Quintana L, Ruiz MA, González D, Sánchez-Valverde F. Anthropometric characteristics and nutrition in a cohort of PAH-deficient patients. *Clin. Nutr.* 2014 Aug;33(4):702-17.

[8] Giovannini M, Verduci E, Salvatici E, Paci S, Riva E. Phenylketonuria: nutritional advances and challenges. *Nutr. Metab* (Lond) 2012 Feb 3;9(1):7.

[9] Bélanger-Quintana A, Burlina A, Harding CO, Muntau AC. Up to date knowledge on different treatment strategies for phenylketonuria. *Mol. Genet. Metab.* 2011;104 Suppl:S19-25.

[10] Kure S, Hou DC, Ohura T, Iwamoto H, Suzuki S, Sugiyama N, Sakamoto O, Fujii K, Matsubara Y, Narisawa K. Tetrahydrobiopterin-responsive phenylalanine hydroxylase deficiency. *J. Pediatr.* 1999 Sep;135(3):375-8.

[11] Couce ML, Bóveda MD, Valerio E, Pérez-Muñuzuri A, Fraga JM. Long-term pharmacological management of phenylketonuria, including patients below the age of 4 years. *JIMD Rep.* 2012;2:91-6.

[12] Cunningham A, Bausell H, Brown M, Chapman M, DeFouw K, Ernst S, McClure J, McCune H, O'Steen D, Pender A, Skrabal J, Wessel A, Jurecki E, Shediac R, Prasad S, Gillis J, Cederbaum S. Recommendations for the use of sapropterin in phenylketonuria. *Mol. Genet. Metab.* 2012 Jul;106(3):269-76.

[13] Burton BK, Bausell H, Katz R, Laduca H, Sullivan C. Sapropterin therapy increases stability of blood phenylalanine levels in patients with BH4-responsive phenylketonuria (PKU). *Mol. Genet. Metab.* 2010 Oct-Nov;101(2-3):110-4.

[14] Singh RH, Quirk ME, Douglas TD, Brauchla MC. BH_4 therapy impacts the nutrition status and intake in children with phenylketonuria: 2-year follow-up. *J. Inherit. Metab. Dis.* 2010;33:689-695.

[15] Lambruschini N, Pérez-Dueñas B, Vilaseca MA, Mas A, Artuch R, Gassió R, Gómez L, Gutiérrez A, Campistol J. Clinical and nutritional evaluation of phenylketonuric patients on tetrahydrobiopterin monotherapy. *Mol. Genet. Metab.* 2005;86 Suppl1:S54-S60.

[16] Aldámiz-Echevarría L, Bueno MA, Couce ML, Lage S, Dalmau J, Vitoria I, Andrade F, Llarena M, Blasco J, Alcalde C, Gil D, García MC, González-Lamuño D, Ruiz M, Ruiz MA, González D, Sánchez-Valverde F. Tetrahydrobiopterin therapy vs phenylalanine-restricted diet: impact on growth in PKU. *Mol. Genet. Metab.* 2013;109:331-338.

[17] Carrascosa Lezcano A, Ferrández Longás A, Yeste Fernández D, García-Dihinx Villanova J, Romo Montejo A, Copil Copil A, Almar Mendoza J, Salcedo Abizanda S, Gussinyé Canadell M, Baguer Mor L. [Spanish cross-sectional growth study 2008. Part I: weight and height values in newborns of 26-42 weeks of gestational age]. *An. Pediatr.* (Barc) 2008 Jun;68(6):544-51.

[18] Carrascosa Lezcano A, Fernández García JM, Fernández Ramos C, Ferrández Longás A, López-Siguero JP, Sánchez González E, Sobradillo Ruiz B, Yeste Fernández D; Grupo Colaborador Español. [Spanish cross-sectional growth study 2008. Part II. Height, weight and body mass index values from birth to adulthood]. *An. Pediatr.* (Barc) 2008 Jun;68(6):552-69.

[19] Dhondt JL, Largillière C, Moreno L, Farriaux JP. Physical growth in patients with phenylketonuria. *J. Inherit. Metab. Dis.* 1995;18:135-7.

[20] Schaefer F, Burgard P, Batzler U, Rupp A, Schmidt H, Gilli G, Bickel H, Bremer HJ. Growth and skeletal maturation in children with phenylketonuria. *Acta. Paediatr.* 1994 May;83:534-41.

[21] Verkerk PH, van Spronsen FJ, Smit GP, Sengers RC. Impaired prenatal and postnatal growth in Dutch patients with phenylketonuria. The National PKU Steering Committee. *Arch. Dis. Child.* 1994;71:114-8.

[22] Belanger-Quintana A, Martínez-Pardo M. Physical development in patients with phenylketonuria on dietary treatment: a retrospective study. *Mol. Genet. Metab.* 2011;104:480-4.

[23] Martínez Pardo M, Bélanger-Quintana A, García Muñoz MJ, Desviat L, Pérez B, Ugarte M. Protocolo de diagnóstico, tratamiento y seguimiento de las hiperfenilalaninemias. Madrid: 2010. (http://www.ae3com.eu/protocolos/protocolo4.pdf)

[24] Spanish Association for Inborn Errors of Metabolism. Clinical practice guideline for the diagnosis, treatment and register of patients with hyperphenylalaninemia in Spain. R. B. Servicios Editoriales, S.L., 2011.

[25] World Health Organization: *WHO Technical Report Series* (no. 935). Singapore; 2007.

[26] Bueno MA, Lage S, Delgado C, Andrade F, Couce ML, González-Lamuño D, Pérez M, Aldámiz-Echevarría L. New evidence for assessing

tetrahydrobiopterin (BH(4)) responsiveness. *Metabolism* 2012;61:1809-1816.

[27] Dobbelaere D, Michaud L, Debrabander A, Vanderbecken S, Gottrand F, Turck D, Farriaux JP. Evaluation of nutritional status and pathophysiology of growth retardation in patients with phenylketonuria. *J. Inherit. Metab. Dis.* 2003;26(1):1-11.

[28] Lässker U, Zschocke J, Blau N, Santer R. Tetrahydrobiopterin responsiveness in phenylketonuria. Two new cases and a review of molecular genetic findings. *J. Inherit. Metab. Dis.* 2002 Feb;25(1):65-70.

[29] Trefz FK, Aulela-Scholz C, Blau N. Successful treatment of phenylketonuria with tetrahydrobiopterin. *Eur. J. Pediatr.* 2001 May;160(5):315.

[30] Arnold GL, Vladutiu CJ, Kirby RS, Blakely EM, Deluca JM. Protein insufficiency and linear growth restriction in phenylketonuria. *J. Pediatr.* 2002 Aug;141(2):243-6.

[31] van Spronsen FJ, Verkerk PH, van Houten M, Smit GP, van der Meer SB, Bakker HD, Sengers RC. Does impaired growth of PKU patients correlate with the strictness of dietary treatment? National Dutch PKU Steering Committee. *Acta. Paediatr.* 1997 Aug;86(8):816-8.

Reviewed by Prof Antonio Baldellou Vázquez
Affiliation: University of Zaragoza (Zaragoza, Spain)
E-mail: baldellou@telefonica.net

In: Phenylalanine
Editor: Darlene Williamson

ISBN: 978-1-63482-501-6

Chapter 3

ROLE OF PHENYLALANINE IN VITICULTURE AND ENOLOGY

P. Santamaría, R. López, J. Portu, L. González-Arenzana, I. López-Alfaro, A. R. Gutiérrez and T. Garde-Cerdán*

Instituto de Ciencias de la Vid y del Vino (Gobierno de La Rioja-CSIC-Universidad de La Rioja), Logroño, Spain

ABSTRACT

Total nitrogen in grapes consists of inorganic and organic forms. Ammonia salts and amino acids are the most prevalent form of total nitrogen by weight in grapes. They constitute, except proline under good nitrogen content conditions, the assimilable nitrogen fraction, which is used by the yeast in the fermentation process.

Moreover, amino acids are also important because they are precursors of a larger number of compounds in grape and wine. Thus, the phenolic compounds are synthesized by the plant through the flavonoid pathway, starting with phenylalanine as a precursor.

This amino acid is also metabolised in the course of the fermentation by the yeasts, producing 2-phenylethanol and 2-phenylethyl acetate, which are important compounds for the wine aromatic quality.

*Instituto de Ciencias de la Vid y del Vino (Gobierno de La Rioja-CSIC-Universidad de La Rioja). Ctra. Mendavia-Logroño NA 134, Km. 90. 26071 Logroño, Spain. Tel: +34 941291833, fax: +34 941291392, e-mail: psantamaria@larioja.org; Teresa.GardeCerdan@gmail.com.

Therefore, the foliar application of this amino acid in the vineyard could modify the phenolic and aromatic composition of grapes and wines. In recent years, some studies have been conducted about different treatments in vineyard that modified berry composition. These treatments have been carried out to improve the phenolic compound content and the aromatic quality of grapes and wines. However, the effect of the foliar application of phenylalanine on the composition of grape and wine has recently begun to be studied. For these reasons, this paper reviews the role of phenylalanine in the fields of viticulture and enology, focus on its influence on the phenolic and aromatic quality of grape and wine.

INTRODUCTION

The must nitrogen composition of grapes affects the growth and metabolism of yeast what is directly related to the fermentation kinetics, to the fermentation completion and can also have an important effect on the aromatic complexity of wines (Bisson, 1991; Rapp andVersini, 1991; Arias-Gil et al., 2007). The nitrogen content of must has been shown to be highly variable, with 60 to 2400 mg/l reported in surveys of Californian juice samples, and includes one inorganic form and various organic forms (OughandAmerine, 1988; Garde-Cerdán et al., 2009; López et al., 2012). In ripe grapes, amino acids generally represent 30–40% of total nitrogen while the remaining is present as ammonium ions, peptides, proteins and nitrate as well as trace amounts of vitamins, nucleotides and amines (Ribéreau-Gayon et al., 2006).

Saccharomyces cerevisiae, the principal yeast of fermentation, preferentially uses simple nitrogen sources such as ammonium ions and free alpha amino nitrogen compounds, present in the form of primary amino acids (MonteiroandBisson 1991; Jiranek et al., 1995). However, proline and hydroxyproline, secondary amino acids, are not metabolised under usual winemaking conditions (BellandHenschke, 2005). The usable nitrogen fraction is referred to as yeast-assimilable nitrogen (YAN). Themicrobial catabolism of amino acids mainly occurs through the activities of five groups of enzymes (i.e., aminotransferases, decarboxylases, dehydratases, lyases and deaminases) which, intracellularly, convert amino acids into a range of volatile and non-volatile compounds, such as α-keto acids, aldehydes, hydroxy acids, higher alcohols, esters and amines, all playing an important role in determining the organoleptic qualities of wine. The volatile compounds, which form part of the aroma profile of wine, constitute the 'fermentation bouquet', whereas non-volatile compounds, such as polyols and carboxylic acids, affect the flavour. In addition, the yeast interacts with a variety of grape secondary metabolites, which have considerable sensory importance (BellandHenschke, 2005). The hydrolysis of glyco-conjugates and cysteine-conjugates can contribute to the varietal character of wine and modification of various grape phenolic compounds can influence wine colour and mouth-feel (Francis andNewton, 2005; Herderichand Smith, 2005). According to different studies, threonine, phenylalanine, alanine and aspartic acid are the amino acids with the most influence on the aromatic compounds obtained in the fermentation process (Hernández-Orte et al., 2006). Thus, the level of 2-phenylethanol and the formation of 2-phenylethyl acetate are closely related to the level of phenylalanine (Hernández-Orte et al., 2002). It is shown that the addition of phenylalanine in different types of must increases the content of2-phenylethanol, 2-phenylethyl acetate (Hernández-Orte et al., 2006; Garde-CerdánandAncín-Azpilicueta, 2008; Trinh et al., 2010; Lee et al., 2013), 2-phenylethyl isobutyrate and 2-phenylethyl hexanoate (Chen et al., 2014). Besides, phenylalanine is considered the first compound involved in stilbenes and flavonoids biosynthesis. Stilbenes and flavonoids are synthesized by the phenylpropanoid/malonate pathway from phenylalanine that, in turn, is converted into cinnamic acid by phenylalanine ammonia lyase (PAL). The consecutive action of cinnamate 4-hydroxylase (C4H) and 4-coumarate CoA ligase (4CL) transform cinnamic acid into p-coumaryl-CoA. Derived compounds, collectively referred to as polyphenols, are originated at this branching point through the action of the enzymes chalcone synthase (CHS) and stilbene synthase (STS) for flavonoids and stilbenes, respectively (Figure 1) (Lijavetzky et al., 2008). Phenolic compounds exert a wide range of biological functions in plants such as antioxidant capacity, protection against UV-light participating in the plant resistance against a range of biotic and abiotic stresses (SoleckaandKacperska, 2003; Takahama, 2004). They also have been reported to be beneficial to human health by contributing to protection against cardiovascular and other chronic diseases and cancer (Cooper, 2004; de la Lastraand Villegas, 2007). In recent years, there have been studies about different vineyard foliar treatments that modified both grapes and wine composition. Thus, it has been demonstrated that the application of different compounds in pre-harvest increased the levels of phenolic compounds and volatile composition of the grapes and resulting wines (Ruiz-García et al., 2013a,b, 2014). Studies about the vine response to the phenylalanine foliar application could be interesting being precursor of several compounds as resveratrol which is a relevant compound with beneficial health properties (Fernández-Mar et al., 2012).

Moreover phenylalanine is also the precursor of phenolic compounds and of 2-phenylethanol, a positive compound for wine aroma with a rose odour descriptor. Additionally, the phenylalanine addition could affect the must nitrogen composition. The increase of YAN concentration in grapes affects the growth and metabolism of yeasts, which is directly related to the fermentation kinetics, to the fermentation completion and, consequently, to the formation of volatile metabolites (BellandHenschke, 2005).

Different Strategies to Modify Phenolic and Aromatic Composition of Grapes and Wines

As protection against pathogens different elicitors and other chemical compounds have been used to induce or enhance plant defence responses. Apart from their effect on activating defence mechanisms, other positive effects have been observed, such as an increase in grape phenolic content and the production of other secondary metabolites. That is the reason because these compounds have been applied in the vineyard to achieve an increase in the volatile and phenolic compounds of grape and wine.

Between these compounds, methyl jasmonate (MeJ) and benzothiadiazole (BTH) have been used to enhance the composition and grape properties. BTH is a functional analogous of salicylic acid a compound that acts as an endogenous signal when plants are primed to resist pathogens (Iriti et al., 2005).

MeJ is a plant volatile compound derived from jasmonic acid (JA). It has similar activity to JA in plants and so is able to activate the enzymes responsible for the biosynthesis of polyphenols (Ruiz-Garcíaand Gómez-Plaza, 2013).

Several studies describe how both treatments and their combination increased the main flavonid compounds in grapes (anthocyanin, flavonol and proanthocyanidin) content of Monastrell grapes (Ruiz-García et al., 2012, 2013a), although the range of the response differed among different clones from the same variety (Ruiz-García et al., 2013b).

Irati et al. (2004) studied the treatment of Merlot grapevines with BTH, observing an increased on anthocyanin and *trans*-resveratrol synthesis. On the other hand, wines achieved from Syrah grapes treated first with methyl jasmonate before harvest and, second with UVC light after harvest, shown 2-fold stilbene concentration than control (Fernández-Marín et al., 2014).

Besides, it has been shown that MeJ is able to induce anthocyanin synthesis (Tassoni et al., 2012) and resveratrol production (Lijavetzky et al., 2008) in grapevine cell suspensions. Belhadj et al. (2008) demonstrated that treatment of grapevine cell suspensions with MeJ in presence of carbohydrates induces a strong stilbene production with an anthocyanin increase.

Also, the influence of application of 24-epibrassinolide (EBR) in the phenolics content of Cabernet Sauvignon and Yan 73 has been studied. EBR is a brassinosteroid which are implicated in plant growth and development. The results obtained shown that the exogenous EBR treatment promote grape ripening and enhance anthocyanins as other phenolics contents and antioxidant capacity in the grape skin (Xi et al., 2013).

The foliar application of abcisic acid, a plant growth regulator involved in different physiological processes, in Monastrell grapes increased the content of skin anthocyanins and tannins, however this increased were not reflected in elaborated wines (Ruiz-García et al., 2013a).

Changes in the production of volatile compounds in grapes due to environmental stress, biological stimulation or foliar application of different nitrogen sources have also been described. Thus, Gómez-Plaza et al. (2012) observed an enhancement in the levels of volatile compounds in grapes sprayed with BTH and MeJ, especially terpenes in benzothiadiazole-treated grapes. Grapes sprayed with a combination of both elicitors led to wines with higher concentration of terpenes that untreated grapes.

However, the composition in higher alcohols and esters were similar in both wines and untreated grapes (Ruiz-García et al., 2014).

In recent years, some studies have been carried out about the influence of foliar application of nitrogen compounds, mainly urea, on the grape composition. It has been shown that foliar urea applications might increase berry YAN and some amino acids in grapes (Lasa et al., 2012; Neilsen et al., 2013) and it improve the aromatic quality of the wine (Dufourcq et al., 2009; Geffroy et al., 2012; Ancín-Azpilicueta, et al., 2013).

PHENYLALANINE FOLIAR APPLICATION TO VINEYARD

In 2012, we began researching the effect of different nitrogen resources applied to canopy of Tempranilo vineyard. These studies have been targeted to analyse the effect of the treatments in the grape aromatic composition (Garde-Cerdán et al., 2014a), in the primary aromatic compounds of grapes (Garde-Cerdán et al., 2015a) and in grape and wine stilbenes (Garde-Cerdán et al.,

2015b). The phenylalanine foliar application was one of the treatments developed. In Table 1 the oenological parameters and YAN of the musts from control and treated samples are displayed. No significant differences were found for any of the physicochemical parameters studied. Regarding the YAN, the samples from PheT treatments showed higher concentrations of this parameter than the control sample did. In all cases, YAN values were found to be above the limit considered by some authors as suitable for the correct development of alcoholic fermentation (Bely et al., 1990).

The effect of PheT application caused an increase in the concentration of the amino acids serine, threonine, phenylalanine, isoleucine, and arginine of treated samples. This made total amino acids composition and total amino acids without proline being higher in treated grapes than in control ones (Figure 2). Amino acids are necessary for yeast metabolism and moreover are precursors of several fermentative volatile compounds (Garde-CerdánandAncín-Azpilicueta, 2008); therefore, nitrogen composition is a key factor in wine quality. As a result, the synthesis of amino acids during the alcoholic fermentation of esters, which are the main compounds for wine aroma being decisive in its organoleptic quality, is related to the must nitrogen composition (Martínez-Gil et al., 2012).

There was a very relevant increase in the phenylalanine concentration in the samples from the treatment with this amino acid, which indicated that the plant assimilated it and transported it to the berry.

Table 1. Physico-chemical parameters and yeast assimilable nitrogen (YAN) of control and treated musts with phenylalanine (PheT)

	Control	PheT
Weight of 100 berries (g)	185±23a	157±16a
Probable alcohol (% v/v)	13.2±0.9a	13.5±0.3a
pH	3.37±0.08a	3.49±0.18a
Total acidity (g/l)*	6.24±0.19b	5.59±0.32a
Tartaric acid (g/l)	5.83±0.08a	5.80±0.17a
Malic acid (g/l)	3.01±0.54a	2.50±0.58a
Potassium (mg/l)	1658±56a	1902±406a
YAN (mg N/l)	169.4±13.9a	267.1±40.6b

All parameters are listed with their standard deviation (n = 3). For each parameter, values with the same letters are not significantly different between the samples ($p \leq 0.05$). *As g/l tartaric acid.From Garde-Cerdán et al. (2014a).

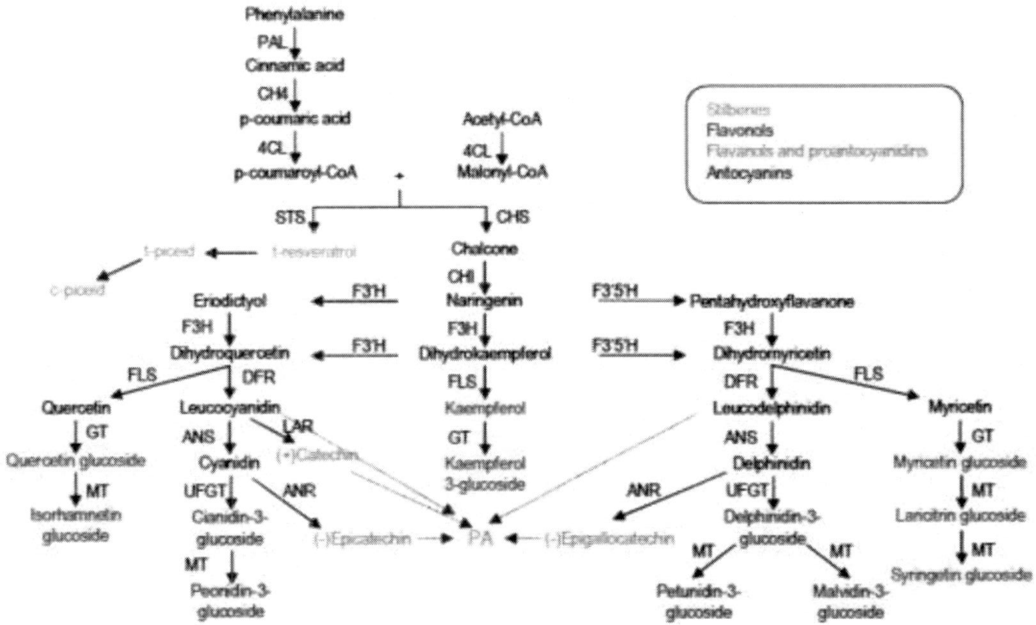

PAL, phenylalanine ammonia-lyase; C4H, cinnamate 4-hydroxylase; 4CL, 4-coumarate CoA ligase; STS, stilbene synthase; CHS, chalcone synthase; CHI, chalconeisomerase; F3H, flavanone-3b-hydroxylase; F3'H, flavonoid 3'-hydroxylase; F3'5'H, flavonoid 3'5'-hydroxylase; DFR, dihydroflavonol-4-reductase; ANS, anthocyanidin synthase; UFGT, UDP-glucose: flavonoid 3-O-glucosyltransferase; FLS, flavonolsyntetase; LAR, leucoanthocyanidinreductasa; ANR, anthocyanidinreductasa; GT, glucosyltransferase; MT, methyltransferase; PA, proanthocyanidins(from Ruiz, 2014).

Figure 1. Schematic diagram of the phenolic biosynthetic pathway.

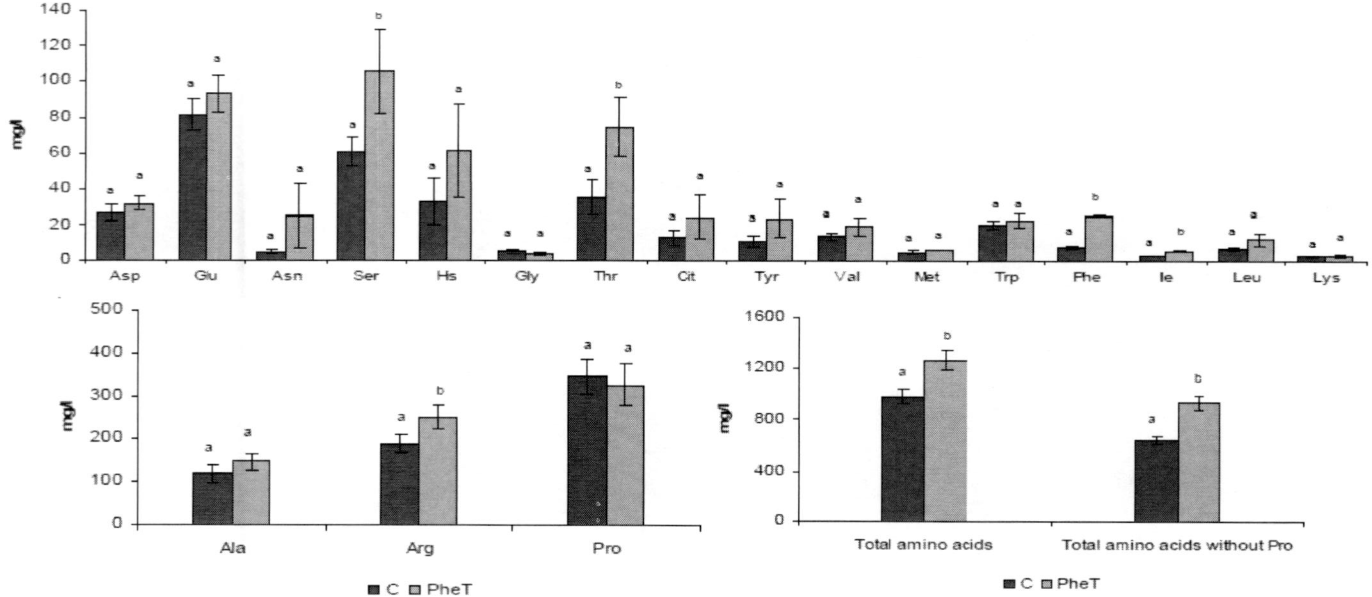

From Garde-Cerdán et al. (2014a).

Figure 2. Amino acid concentration (mg/l) in the control (C) and treated with phenylalanine (PheT) samples. All parameters are listed with the standard deviation (n = 3). For each parameter, different letters indicate differences between the samples ($p \leq 0.05$).

As mentioned above, phenylalanine is the precursor of phenolic compounds and 2-phenylethanol, which are very important compounds for the quality and the health properties of must and wine.

Regarding the effect of this nitrogen treatment (PheT) on the grape aroma composition, the results for the control sample and for the treatment ones are shown in Table 2. Terpenoids play a significant role on the varietal wine aroma, contributing to the floral and citrus character of wines (Ribéreau-Gayon et al., 2006). The terpenoids synthesis decreased with PheT treatment applied in the vineyard, thus total terpenoids were lower in the samples from these treated vines than in those from the control vines. Taking into account data about each one of the terpenoids, it was observed that citronellol, nerol, neral, geraniol, γ-geraniol and p-cymene were unaffected by the treatment, since no significant differences were observed respect to the control, while PheT application decreased the content of linalool, α-terpineol and geranyl acetone. Regarding total C_{13}norisoprenoids, the treated samples showed no significant differences with the control (Table 2). This same behavior was described for all the individual C_{13}norisoprenoids. Among these compounds, β-damascenone and β-ionone are the most important ones as they contribute to the flavor and odor, due to their low perception thresholds, providing violet and rose aromas (Kotseridis et al., 1999; Sefton et al., 2011). The phenylalanine foliar application in Tempranillo vineyard did not affected the formation of esters identified in the grapes, since no significantdifferences were observed between the control sample and treated samples (Table 2). Although esters are highly relevant for wine aroma, they are formed in high quantity as a consequence of alcoholic fermentation (Jackson, 2008). Benzenoid compounds confer desirable aroma to wine, particularity 2-phenylethanol, with a rose aroma descriptor (Etiévant, 1991; Furdíková et al., 2014). Among the results obtained in this study about benzenoid compounds, those observed for phenylalanine treatment were noteworthy. An increase of 178 % of 2-phenylethanol+2-phenylethanal content compared to the control sample was detected (Table 2). This meant that the plant transformed the phenylalanine applied over the leaves in both 2-phenylethanol and 2-phenylethanal. Due to the plant metabolism after this treatment, total benzenoids were significantly higher for the phenylalanine treatment than those obtained for the control one. The foliar treatments with phenylalanine decreased total C6 compounds in the samples in relation to the control. This might be positive for the aroma, because these compounds at high levels can provide undesirable herbaceous flavors to the product (Zalacain et al., 2007; Garde-Cerdán et al., 2013).

Table 2. Volatile compound area (x10^6) in control and treated must with phenylalanine (PheT). All parameters are listed with their standard deviation (n = 3)

	Control	PheT
Terpenoids		
Linalool	1.04±0.11b	0.55±0.01a
α-Terpineol	0.63±0.03b	0.37±0.01a
Citronellol	0.05±0.01a	0.06±0.01a
Nerol	0.17±0.03a	0.12±0.02a
Neral	0.06±0.01a	0.05±0.01a
Geraniol	0.47±0.13a	0.54±0.35a
Geranyl acetone	2.69±0.05b	0.85±0.25a
γ-Geraniol	0.09±0.01a	0.05±0.02a
p-Cymene	0.17±0.06a	0.11±0.01a
Total	*5.35±0.25b*	*2.71±0.62a*
C_{13}norisoprenoids		
(E)-β-Damascenone	20.46±2.80a	13.80±3.96a
(Z)-β-Damascenone	1.40±0.17a	1.02±0.32a
β-Ionone	1.81±0.34a	1.61±0.13a
β-Cyclocitral	0.34±0.06a	0.30±0.01a
Methyl jasmonate	0.29±0.12a	0.23±0.12a
Total	*24.30±2.69a*	*16.94±4.29a*
Esters		
2-Hexen-1-ol acetate	0.74±0.27a	0.63±0.06a
Hexyl acetate	5.57±0.61a	5.02±3.82a
Methyl salicylate	0.11±0.02a	0.12±0.07a
Total	*6.42±0.91a*	*5.76±3.69a*
Benzenoids		
2-Phenylethanol	10.13±3.74a	29.41±4.02b
2-Phenylethanal	2.82±0.35a	6.53±0.70b
Benzyl alcohol	0.91±0.12b	0.64±0.06a
Total	*13.85±4.21a*	*36.57±4.79b*
C6 compounds		
n-Hexanol	650.20±72.21a	486.91±167.79a
(Z)-3-Hexen-1-ol	25.80±3.96a	17.03±4.56a
(E)-2-Hexen-1-ol	75.01±14.04a	90.79±37.45a
Hexanal	392.66±57.36b	190.70±63.91a
(E)-2-Hexenal	112.09±2.40a	81.65±13.37a
Total	*1255.76±22.51b*	*867.08±84.37a*
*Total positive compounds**	*49.93±7.56a*	*61.98±4.81b*
Total compounds	*1305.68±30.07b*	*929.07±89.18a*

For each compound, different letters indicate differences between the samples ($p \leq 0.05$). *Total positive compounds=terpenoids+C13 norisoprenoids+esters+benzenoids. From Garde-Cerdán et al. (2015a).

Taking into account the sum of positive compounds for aroma (terpenoids, C_{13}norisoprenoids, esters, and benzenoids), the treatment with phenylalanine favored the presence of these compounds, mainly due to the increase of the synthesis by the plant of 2-phenylethanol and 2-phenylethanal.

Figure 3 shows the results of the concentration of stilbenes in both control must and musts from the phenylalanine (PheT) treatment, as well as the results described for the wines. The total stilbenes quantity, *trans*-resveratrol and *cis*-piceid determined for must and wines obtained from PheT treatment did not demonstrated significant differences with the control ones.

Spite no detecting significant differences in the *trans*-piceid content in musts; the concentration of that compound in wine samples from PheT treatment was higher than in control samples, being the majoritarystilben in grapes (Romero-Pérez et al., 1999; López-Alfaro et al., 2013). Grapes and wines are considered the most important stilbene sources (Mattivi et al., 1995), beingrelevant compounds, particularly the *trans*-resveratrol, due to its widely described health properties (Fernández-Mar et al., 2012).

Following with this research line, Chassy et al. (2014) carried out a study in which Cabernet Sauvignon grape clusters were treated in unripe green grapes and ripening red grapes with $^{13}C_6$-phenylalanine. In spite of the quantity of this isotope was not specified, results demonstrate a massive incorporation of $^{13}C_6$-phenylalanine into anthocyanins, probably because of their dominant accumulation during grape maturation (Portu et al., 2015). Furthermore, authors suggested the presence of a previously unreported pool of substrate in the phenolic pathway in grapevines, which should be the aim of future research.

Phenylalanine Metabolism during the Alcoholic Fermentation

Phenylalanine is present in the whole grape and/or juice at concentrations from 2.8 to 138 mg/L (Bell andHenschke, 2005) being considered as one of the yeast's favorite (Martínez-Moreno et al., 2014).

During the earlier alcoholic fermentation stages, phenylalanine goes being metabolised by yeast (Hernández-Orte et al., 2006; Arias-Gil et al., 2007; Garde-Cerdán et al., 2007).

The metabolic rate of phenylalanine is directly linked with its initial concentration in grapes (Garde-Cerdán et al., 2011; Portu et al., 2014).

Besides being an important nitrogen source for the correct development of alcoholic fermentation, phenylalanine is the amino acid precursor of 2-phenylethanol y de 2-phenylethyl acetate, really important compounds for aromatic quality of wines. In terms of this, Garde-Cerdán and Ancín-Azpilicueta (2008) described that the synthesis of those aromatic compounds, was directly proporcional to the quantity of amino acids added to a Mazuelo must initially poor in nitrogen compounds.

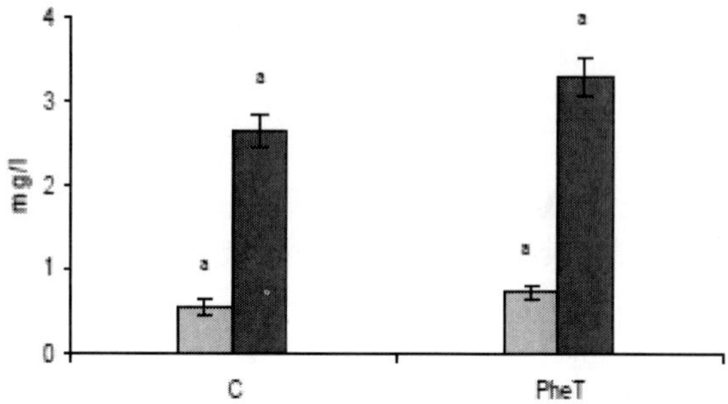

a

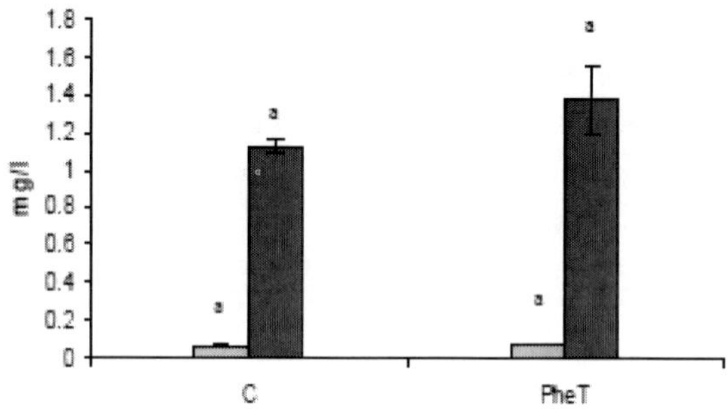

b

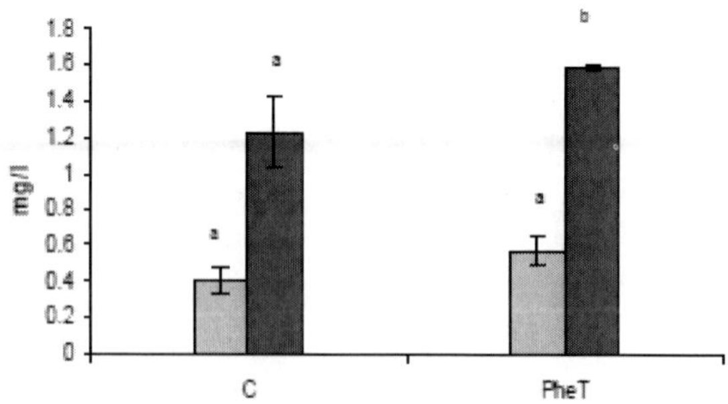

c

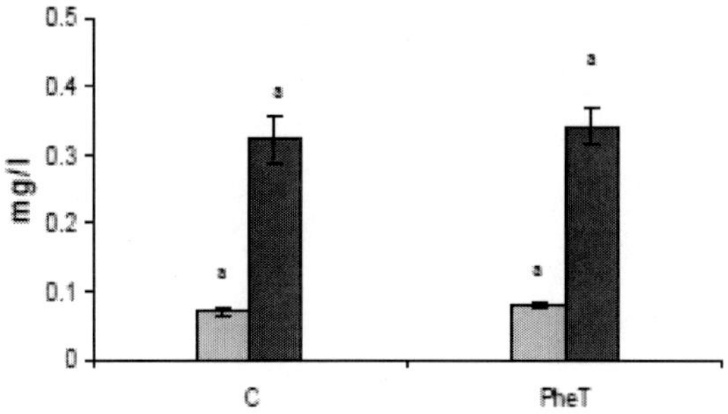

d
[a]Totalstilbenes.
[b]*trans*-Resveratrol.
[c]*trans*-Piceid.
[d]*cis*-Piceid.
From Garde-Cerdán et al. (2015b).

Figure 3. Stilbene concentration (mg/l) in the control (C) and treated with phenylalanine (PheT) musts and wines. All parameters are listed with the standard deviation (n = 3). At each sampling of winemaking process, different letters indicate differences between the samples ($p \leq 0.05$).

Similar results were described by Hernández-Orte et al. (2006) who determined that adding phenylanaline, aspartic acid, threonine, and alanine to a Merlot must provided a greater synthesis of 2-phenylethanol during the fermentation. Morevover, Trinh et al. (2010) also found significant differences in the 2-phenylethanol and its corresponding acetate synthesis after adding isoleucine and phenylalanine to the must. Contrarily, Martínez-Gil et al. (2012) did not described a quite clear relation between the inicial phenylalanine concentration in four varieties musts and the 2-phenylethanol and 2-phenylethyl acetate synthesis during the alcoholic fermentation.

Furthermore, it is widely known the existence of amino acids precursor of biogenic amines. These amines are aliphatic, aromatic or heterocyclic organic bases of low molecular mass that occur in plants and in fermented foods. These compounds are undesirable in all foods and beverages because if consumed at an excessively high concentration, they may induce headaches, respiratory distress, heart palpitation, hyper-hypotension, and several allergenic disorders in man (Silla-Santos, 1996). These amines are synthesized by microbial decarboxylation of the corresponding amino acids (Halásk et al., 1994) mainly during malolactic fermentation (Ancín-Azpilicueta et al., 2008) because during alcoholic fermentation the biogenic amines synthesis is not really usual (Garde-Cerdán et al., 2008).

Phenylalanine is precursor of phenylethylamine that is one of the most commonly detected in wines (López et al., 2012; Garde-Cerdán et al., 2014b) though Martínez-Pinilla et al. (2013) described a negative correlation of this amine with phenylalanine. In alcoholic beverages, 3 mg/L of this amine can cause negative physiological effects (Soufleros et al., 1998). Moreover, the amine concentration affects probably the exportability of wines as some countries have set maximum limits for the content of these compounds (Beneduce et al., 2010).

Conclusion

The phenylalanine is a remarkable amino acid in the wine quality, significantly influencing both its aromatic and phenolic composition and its healthy properties, being the precursor of the phenolic compounds in their plant synthesis metabolic pathway. Therefore, those treatments in the vineyard that encourage the phenylalanine synthesis positively influence in the grape composition, and as a consequence, in the final wine. Additionally, this amino acid is a nitrogen source for yeast metabolisms what benefits the adequate development of the fermentation and is also the precursor of significant aromatic compounds in the final global wine aroma.

Acknowledgments

J. P. thanks to Gobierno de La Rioja for the formation grant. T. G.-C., L. G.-A., and I. L.-A. also wish to thank the InstitutoNacional de Investigación y TecnologíaAgraria y Alimentaria (INIA)-Gobierno de La Rioja and FEDER of the European Community for the doctoral and pre-doctoral support.

References

Ancín-Azpilicueta, C., González-Marco, A., Jiménez-Moreno, N. (2008). Current knowledge about the presence of amines in wine. *Critical Reviews in Food Science and Nutrition*, 48, 257-275.

Ancín-Azpilicueta, C., Nieto-Rojo, R., Gómez-Cordón, J. (2013). Effect of foliar urea fertilisation on volatile compounds in Tempranillo wine. *Journal of the Science of Food and Agriculture*, 93, 1485–1491.

Arias-Gil, M., Garde-Cerdán, T., Ancín-Azpilicueta, C. (2007). Influence of addition of ammonium and different amino acid concentrations on nitrogen metabolism in spontaneous must fermentation. *Food Chemistry*, 103, 1312-1318.

Belhadj, A., Telef, N., Saigne, C., Cluzet, S., Barrieu, F., Hamdi, S., Mérillon, J.-M. (2008). Effect of methyl jasmonate in combination with carbohydrates on gene expression of PR proteins, stilbene and anthocyanin accumulation in grapevine cell cultures. *Plant Physiology and Biochemistry*, 46, 493-499.

Bell, S.-J., Henschke, P. A. (2005). Implications of nitrogen nutrition for grapes, fermentation and wine. *Australian Journal of Grape and Wine Research*, 11, 242-295.

Bely, M., Sablayrolles, J. M., Barre, P. (1990). Automatic detection of assimilable nitrogen deficiencies during alcoholic fermentation in enological conditions. *Journal of Fermentation and Bioengineering*, 70, 246-252.

Beneduce, L., Romano, A., Capozzi, V., Lucas, P., Barnavon, L., Bach, B., Vuchot, P., Grieco, F., Spano, G. (2010). Biogenic amine in wines. *Annals of Microbiology*, 60, 573-578.

Bisson, L. F. (1991). Influence of nitrogen on yeast and fermentation of grapes. In:*Proceedings of the International Symposium on Nitrogen in Grapes and Wine* (pp. 78-89), Rantz, J. (Ed.), American Society for Enology and Viticulture, Davis.

Chassy, A. W., Adams, D. O., Waterhouse, A. L. (2014). Tracing phenolic metabolism in *Vitisvinifera* berries with $^{13}C_6$-phenylalanine: Implication of an unidentified intermediate reservoir. *Journal of Agricultural and Food Chemistry*, 62, 2321-2326.

Chen, D., Chia, J. Y., Liu, S.-Q. (2014). Impact of addition of aromatic amino acids on non-volatile and volatile compounds in lychee wine fermented with *Saccharomyces cerevisiae*MERIT.ferm. *International Journal of Food Microbiology*, 170, 12–20.

Cooper, K. A., Chopra, M., Thurnham, D. I. (2004). Wine polyphenols and promotion of cardiac health. *Nutrition Research Reviews*, 17, 111–129.

de la Lastra, C.A., Villegas, I. (2007). Resveratrol as an antioxidant and prooxidant agent: mechanisms and clinical implications. *Biochemical Society Transactions*, 35, 1156-1160.

Dufourcq, T., Charrier, F., Poupault, P., Schneider, R., Gontier, L., Serrano, E. (2009). Foliar spraying of nitrogen and sulfur at veraison: a viticultural technique to improve aromatic composition of white and rosés wines. In:*Proceedings of the 16th International GiESCO Symposium* (pp. 379–383), Department of Viticulture and Enology, Davis.

Etiévant, P. X. (1991). Wine. In:*Volatile Compounds in Foods and Beverages* (pp. 483-546), Maarse, H. (Ed.), Marcel Dekker, New York.

Fernández-Marín, M. A., Puertas, B., Guerrero, R. F., García-Parrilla, M. C., Cantos-Villar, E. (2014). Preharvest methyl jasmonate and postharvest UVC treatments: increasing stilbenes in wine. *Journal of FoodScience*, 79, 310-317.

Fernández-Mar, M. I., Mateos, R., García-Parrilla, M.C., Puertas, B., Cantos-Villar, E. (2012). Bioactive compounds in wine: Resveratrol, hydroxytyrosol and melatonin: A review. *Food Chemistry*, 130, 797-813.

Francis, I. L., Newton, J. L. (2005). Determining wine aroma from compositional data. *Australian Journal of Grape and Wine Research*, 11, 114–126.

Furdíková, K., Makyšová, K., Ďurčanská, K., Špánik, I., Malík, F. (2014). Influence of yeast strain on aromatic profile of Gewürztraminer wine. *LWT-Food Science and Technology*, 59, 256-262.

Garde-Cerdán, T., Marsellés-Fontanet, A. R., Arias-Gil, M., Martín-Belloso, O., Ancín-Azpilicueta, C. (2007). Influence of SO_2 on the consumption of nitrogen compounds through alcoholic fermentation of must sterilized by pulsed electric fields. *Food Chemistry*, 103, 771-777.

Garde-Cerdán, T., Ancín-Azpilicueta, C. (2008). Effect of the addition of different quantities of amino acids to nitrogen-deficient must on the formation of esters, alcohols, and acids during wine alcoholic fermentation. *LWT-Food Science and Technology*, 41, 501-510.

Garde-Cerdán, T., Arias-Gil, M., Marsellés-Fontanet, A. R., Salinas, M. R., Ancín-Azpilicueta, C., Martín-Belloso, O. (2008). Study of the alcoholic fermentation of must stabilized by pulsed electric fields-Effect of SO_2. In:*Progress in Food Chemistry* (pp. 73-104), Koeffer, E. N. (Ed.), Nova Science Publishers, Inc., New York.

Garde-Cerdán, T., Lorenzo, C., Lara, J. F., Pardo, F., Ancín-Azpilicueta, C., Salinas, M. R. (2009). Study of the evolution of nitrogen compounds during grape ripening. Application to differentiate grape varieties and cultivated systems. *Journal of Agricultural and Food Chemistry*, 57, 2410-2419.

Garde-Cerdán, T., Martínez-Gil, A. M., Lorenzo, C., Lara, J. F., Pardo, F., Salinas, M. R. (2011). Implications of nitrogen compounds during alcoholic fermentation from some grape varieties at different maturation stages and cultivation systems. *FoodChemistry*, 124, 106-116.

Garde-Cerdán, T., González-Arenzana, L., López, N., López, R., Santamaría, P., López-Alfaro, I. (2013). Effect of different pulsed electric field treatments on the volatile composition of Graciano, Tempranillo and Grenache grape varieties. *Innovative Food Science and Emerging Technologies*, 20, 91-99.

Garde-Cerdán, T., López, R., Portu, J., González-Arenzana, L., López-Alfaro, I., Santamaría, P. (2014a). Study of the effects of proline, phenylalanine, and urea foliar application to Tempranillo vineyards on grape amino acid content. Comparison with commercial nitrogen fertilizers. *Food Chemistry*, 163, 136-141.

Garde-Cerdán, T., López, R., Garijo, P., González-Arenzana, L., Gutiérrez, A. R., López-Alfaro, I., Santamaría, P. (2014b). Application of colloidal silver versus sulfur dioxide during vinification and storage of Tempranillo red wines. *Australian Journal of Grape and Wine Research*, 20, 51-61.

Garde-Cerdán, T., Santamaría, P., Rubio-Bretón, P., González-Arenzana, L., López-Alfaro, I., López, R. (2015a). Foliar application of proline, phenylalanine, and urea to Tempranillo vines: Effect on grape volatile composition and comparison with the use of commercial nitrogen fertilizers. *LWT-Food Science and Technology*, 60, 684-689.

Garde-Cerdán, T., Portu, J., González-Arenzana, L., López, R., Santamaría, P. (2015b). Impact of proline, phenylalanine, urea and nitrogen fertilizers foliar treatments to vine on grape and wine stilbene. *American Journal of Enology and Viticulture*, under review.

Geffroy, O., Dufourcq, T., López, R., Serrano, E., Gracia-Moreno, E., Cacho, J., Ferreira, F. (2012). Impact de pulvérisations foliaires azotées réalisées à la véraison sur les caractéristiques aromatiques des vins rouges. In:*Actas del Coloquio Internacional sobre los Aromas del Vino.* Proyecto VINAROMAS (pp. 87–89), Zaragoza, Spain.

Gómez-Plaza, E., Mestre-Ortuño, L., Ruiz-García, Y., Fernández-Fernández, J. I., López-Roca, J.M. (2012). Effect of benzothiadiazole and methyl jasmonate on the volatile compound composition of *Vitisvinifera*L. Monastrell grapes and wines. *American Journal of Enology and Viticulture*, 634, 694-401.

Halász, A., Baráth, A., Simon-Sarkadi, L. S., Holzapfel, W. (1994). Biogenic amines and their production by microorganisms in food. *Trends in Food Science and Technology*, 3, 42-49.

Herderich, M. J., Smith, P. A. (2005). Analysis of grape and wine tannins: Methods, applications and challenges. *Australian Journal of Grape and Wine Research*, 11, 205–214.

Hernández-Orte, P., Cacho, J., Ferreira, V. (2002). Relationship between varietal amino acid profile of grapes and wine aromatic composition. Experiments with model solutions and chemometric study. *Journal of Agricultural and Food Chemistry*, 50, 2891–2899.

Hernández-Orte, P., Ibarz, M. J., Cacho, J., Ferreira, V. (2006). Addition of amino acids to grape juice of the Merlot variety: Effect on amino acid uptake and aroma generation during alcoholic fermentation. *Food Chemistry*, 98, 300-310.

Iriti, M., Rossoni, M., Borgo, M., Faoro, F. (2004). Benzothiadiazole enhances resveratrol and anthocyanin biosynthesis in grapevine, meanwhile improving resistance to *Botrytis cinerea*. *Journal of Agricultural and Food Chemistry*, 52, 4406−4413.

Iriti, M., Rossoni, M., Borgo, M., Ferrara, L., Faoro, F. (2005). Induction of resistance to graymold with benzothiadiazole modifies amino acid profile and increases proanthocyanidins in grape: Primary versus secondary metabolism. *Journal of Agricultural and Food Chemistry*, 53, 9133-9139.

Jackson, R.S. (2008). *Wine Science. Principles and Applications*. Academic Press, New York.

Jiranek, V., Langridge, P., Henschke, P. A. (1995). Amino acid and ammonium utilization by *Saccharomyces cerevisiae* wine yeasts from a chemically defined medium. *American Journal of Enology and Viticulture*,46, 75–83.

Kotseridis, Y., Baumes, R. L., Bertrand, A., Skouroumounis, G. K. (1999). Quantitative determination of β-ionone in red wines and grapes of Bordeaux using a stable isotope dilution assay. *Journal of Chromatography A*, 848, 317-325.

Lasa, B., Menendez, S., Sagastizabal, K., Cervantes, M. E. C., Irigoyen, I., Muro, J., Aparicio-Tejo, P. M., Ariz, I. (2012). Foliar application of urea to "Sauvignon Blanc" and "Merlot" vines: doses and time of application. *Plant Growth Regulation*, 67, 73–81.

Lee, P.-R., Toh, M., Yu, B., Curran, P., Liu, S.-Q. (2013). Manipulation of volatile compound transformation in durian wine by nitrogen supplementation. *International Journal of Food Science and Technology*, 48, 650–662.

Lijavetzky, D., Almagro, L., Belchi-Navarro, S., Martínez-Zapater, J. M., Bru, R., Pedreño, M. A. (2008). Synergistic effect of methyljasmonate and cyclodextrin on stilbene biosynthesis pathway gene expression and resveratrol production in Monastrell grapevine cell cultures. *BMC Research Notes*, 1, 132.

López, R., Tenorio, C., Gutiérrez, A. R., Garde-Cerdán, T., Garijo, P., González-Arenzana, L., López-Alfaro, I., Santamaría, P. (2012). Elaboration of Tempranillo wines at two different pHs. Influence on biogenic amine contents. *Food Control*, 25, 583-590.

López-Alfaro, I., González-Arenzana, L., López, N., Santamaría, P., López, R., Garde-Cerdán, T. (2013). Pulsed electric field treatment enhanced stilbene content in Graciano, Tempranillo and Grenache grape varieties. *FoodChemistry*, 141, 3759-3765.

Martínez-Gil, A. M., Garde-Cerdán, T., Lorenzo, C., Lara, J. F., Pardo, F., Salinas, M. R. (2012). Volatile compounds formation in alcoholic fermentation from grapes collected at 2 maturation stages: influence of nitrogen compounds and grape variety. *Journal of Food Science*, 77, C71-C79.

Martínez-Moreno, R., Quirós, M., Morales, P., Gonzalez, R. (2014). New insights into the advantages of ammonium as a winemaking nutrient. *International Journal of Food Microbiology*, 177, 128-135.

Martínez-Pinilla, O., Guadalupe, Z., Hernández, Z., Ayestarán, B. (2013). Amino acids and biogenic amines in red varietal wines: the role of grape variety, malolactic fermentation and vintage. *European Food Research and Technology*, 237, 887-895.

Mattivi, F., Reniero, F., Korhammer, S. (1995). Isolation, characterization, and evolution in red wine vinification of resveratrol monomers. *Journal of Agricultural and Food Chemistry*, 43, 1820-1823.

Monteiro, F. F., Bisson, L. F. (1991). Amino acid utilisation and urea formation during vinification fermentations. *American Journal of Enology and Viticulture*, 42, 199–208.

Neilsen, G. H., Neilsen, D., Hannam, K., Millard, P., Midwood, A. (2013). Foliar urea applications can increase berry yeast-assimilable N in wine grapes (*Vitisvinifera* L.). *ActaHorticulturae*, 984, 427-434.

Ough, C. S., Amerine, M. A. (1988). *Methods for analysis of musts and wines* (Second edition), Wiley-Interscience, New York.

Portu, J., López, R., López-Alfaro, I., González-Arenzana, L., Santamaría, P., Garde-Cerdán, T. (2014). Amino acid content in red wines obtained from grapevine nitrogen foliar treatments: Consumption during the alcoholic fermentation. *WineStudies*, 3, 2-6.

Portu, J., López-Giral, N., López, R., González-Arenzana, L., González-Ferrero, C., López-Alfaro, I., Santamaría, P., Garde-Cerdán, T. (2015). Different tools to enhance grape and wine anthocyanin content. In:*Handbook of Anthocyanins-Food Sources, Chemical Applications and Health Benefits* (pp. 51-88), Warner, L. M. (Ed.), Nova Science Publishers, Inc., New York.

Rapp, A., Versini, G. (1991). Influence of nitrogen compounds in grapes on aroma compounds in wine. In:*Proceedings of the International Symposium on Nitrogen in Grapes and Wine* (pp. 156-164), Rantz, J. (Ed.), American Society for Enology and Viticulture, Davis.

Ribéreau-Gayon, P., Glories, Y., Maujean, A., Dubourdieu D. (2006). *Handbook of Enology. Volume 2. The Chemistry of Wine Stabilization and Treatments*, Jonh Wiley and Sons, Ltd., Chichester.

Romero-Pérez, A. I., Ibern-Gómez, M., Lamuela-Raventós, R. M., de la Torre-Boronat, M. C. (1999). Piceid, the major resveratrol derivative in grape juices. *Journal of Agricultural and Food Chemistry*, 47, 1533-1536.

Ruiz-García, Y., Romero-Cascales, I., Gil-Muñoz, R., Fernández-Fernández, J. I., López-Roca, J. M., Gómez-Plaza, E. (2012). Improving grape phenolic content and wine chromatic characteristics through the use of two different elicitors: methyl jasmonate versus benzothiadiazole. *Journal of Agricultural and Food Chemistry*, 60, 1283−1290.

Ruiz-García, Y., Gil-Muñoz, R., López-Roca, J. M., Martínez-Cutillas, A., Romero-Cascales, I., Gómez-Plaza, E. (2013a). Increasing the phenolic compound content of grapes by preharvest application of abcisic acid and a combination of methyl jasmonate and benzothiadiazole. *Journal of Agricultural and Food Chemistry*, 61, 3978−3983.

Ruiz-García, Y., Gómez-Plaza, E. (2013). Elicitors: A tool for improving fruit phenolic content. *Agriculture*, 3, 33-52.

Ruiz-García, Y., Romero-Cascales, I., Bautista-Ortín, A. B., Gil-Muñoz, R., Martínez-Cutillas, A., Gómez-Plaza, E. (2013b). Increasing bioactive phenolic compounds in grapes; response of six Monastrell grape clones to benzothiadiazole and methyl jasmonate treatments. *American Journal of Enology and Viticulture*, 64, 459-465.

Ruiz-García, Y., López-Roca, J. M., Bautista-Ortín, A.B., Gil-Muñoz, R., Gómez-Plaza, E. (2014). Effect of combined use of benzothiadiazole and methyl jasmonate on volatile compounds of Monastrell wines. *American Journal of Enology and Viticulture*, 65, 238-243.

Ruiz-García, Y. (2014). Elicitores: una herramienta para incrementar el color y el aroma de uvas y vinos. *Doctoral Thesis*, Universidad de Murcia.

Sefton, M. A., Skouroumounis, G. K., Elsey, G. M., Taylor, D. K. (2011). Occurrence, sensory impact, formation, and fate of damascenone in grapes, wines, and other foods and beverages. *Journal of Agricultural and Food Chemistry*, 59, 9717-9746.

Silla-Santos, M. H. (1996). Biogenic amines: their importance in foods. *International Journal of Food Microbiology*, 29, 213-231.

Solecka, D., Kacperska, A. (2003). Phenylpropanoid deficiency affects the course of plant acclimation to cold. *PhysiologiaPlantarum*, 119, 253–262.

Soufleros, E., Barrios, M. L., Bertrand, A. (1998). Correlation between the content of biogenic amines and other wine compounds. *American Journal of Enology and Viticulture*, 49, 266-269.

Takahama, U. (2004). Oxidation of vacuolar and apoplastic phenolic substrates by peroxidase. *Phytochemistry Reviews*, 3, 207–219.

Tassoni, A., Durante, L., Ferri, M. (2012). Combined elicitation of methyl-jasmonate and red light on stilbene and anthocyanin biosynthesis. *Journal of Plant Physiology*, 169, 775-781.

Trinh, T. T. T., Woon, W. V., Yu, B., Curran, P., Liu, S. Q. (2010). Effect of L-isoleucine and L-phenylalanine addition on aroma compound formation during longan juice fermentation by a co-culture of *Saccharomyces cerevisiae* and *Williopsissaturnus*. *South African Journal for Enology and Viticulture*, 31, 116-124.

Xi, Z.-m., Zhang, Z.-w., Huo, S.-s., Luan, L-y., Gao, X., Maa, L.-n., Fang, Y.-l. (2013). Regulating the secondary metabolism in grape berry using exogenous 24-epibrassinolide for enhanced phenolics content and antioxidant capacity. *Food Chemistry*, 141, 3056–3065.

Zalacain, A., Marín, J., Alonso, G. L., Salinas, M. R. (2007). Analysis of wine primary aroma compounds by stir bar sorptive extraction. *Talanta*, 71, 1610-1615.

In: Phenylalanine
Editor: Darlene Williamson
ISBN: 978-1-63482-501-6

Chapter 4

PENETRATION OF L-PHENYLALANINE AND AMINO ACIDS INTO HAIR KERATIN FIBERS

Akio Kuzuhara*
Research and Development Department, Sunny-Place Co., Ltd.,
Kuramae, Taito-ku, Tokyo, Japan

ABSTRACT

In order to investigate the penetration of amino acids and peptides into hair keratin fibers, cross-sectional samples of virgin white human hair treated amino acids and peptides were prepared. We have developed a new method for analyzing the penetration of amino acids and peptides into hair keratin fibers using optical microscopy and dyeing technique. The hydrophilic amino acids and peptides did not penetrate at all, while the hydrophobic amino acids, such as L-phenylalanine (Phe) penetrated into the virgin white human hair. In addition, the cross-sectional samples at various depths of virgin white human hair treated with Phe was directly analyzed without isolating the cuticle and cortex using Raman Spectroscopy. The Phe hydrophobic amino acid content at various depths of the virgin white human hair increased by performing the Phe treatment (at 50°C for 16 h), indicating that Phe deeply penetrated into the virgin human hair.

* Corresponding to: A. Kuzuhara; email address: spu62vm9@voice.ocn.ne.jp, kuzuhara_sunnyplace@xqd.biglobe.ne.jp.

INTRODUCTION

Keratin fibers, such as wool and hair, are composed of three cell layers (the cuticle, the cortex, and sometimes, the medulla). The hierarchical structure of a fine wool fiber is shown in Figure 1. The cortex consists of spindle-shaped macrofibrils that have two main structures, the keratin and the keratin associated protein (KAP), which are distinguished by their structures and amino acid compositions [1-6]. The keratin is a fibrous protein which is mainly composed of the α-helical proteins of a low cystine content. These structures are aligned along the fiber axis and embedded in an amorphous KAP with a high cystine content consisting of disulfide (–SS–) groups. The –SS– groups form the cross-linkages in keratin fibers, and contribute to physical and mechanical properties as well as structural stability.

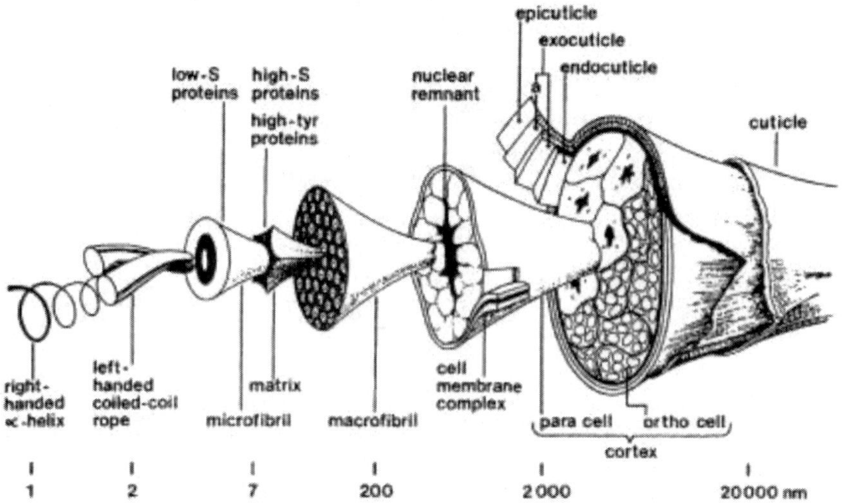

Figure 1. The hierarchical structure of a fine wool fiber [4].

The α-helical proteins existing in the keratin form two-strand helical ropes consisting of right-handed α-helices. The amino acid residue sequence of the purely helical segments forms into heptad repeats $(a\text{-}b\text{-}c\text{-}d\text{-}e\text{-}f\text{-}g)_n$ in which the residues "a" and "d" (alanine, leucine, isoleucine, and tyrosine etc.) are apolar, and "e" and "g" (asparatic acid, glutamic acid, lysine, arginine, histigine) oppositely charged (Figure 2) [6]. The helical structures form into two left-handed coiled-coil ropes stabilized by the hydrophobic and Coulombic interactions as indicated (Figure 2). In the presence of water molecules, the hydrophobic interactions produced between the residues "a" and "d" by the water network "protect" the Coulombic interactions between the residues "e" and "g" from the approach of water molecules. The Coulombic interactions reinforce the binding of the two-strand helical rope.

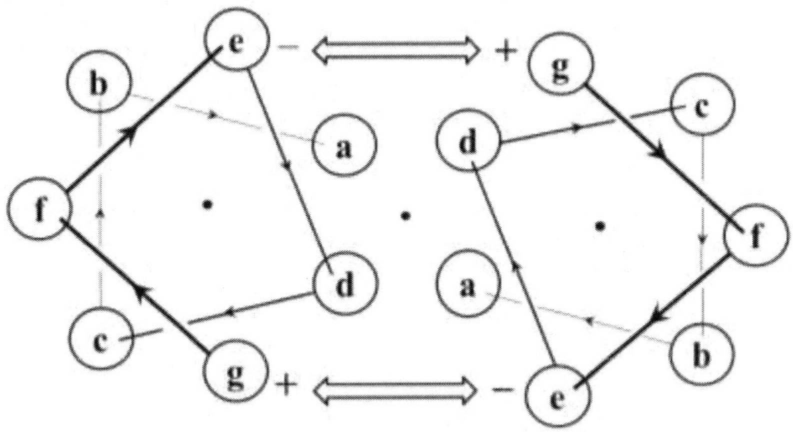

Figure 2. The α-helices form into two-strand ropes with interactions between the helices a result of the heptad repeats of residues $(a\text{-}b\text{-}c\text{-}d\text{-}e\text{-}f\text{-}g)_n$ [6]. The side chains of the residues "a" and "d" are apolar, and "e" and "g" oppositely charged. The results is a combined hydrophobic and Coulombic interaction between the α-helices. The figure shows two heptad segments of each α-helix viewed in the direction of the axes of the helices.

Therefore, it is important to obtain information about –SS– groups and amino acids, when investigating the influence of chemical modification on the structure of keratin fibers.

Hydrolyzed proteins, which are derived from eggwhites, keratin, silk fibroin, soybean, wheat etc., are widely used as an effective ingredient in hair care products, such as shampoo and conditioner in order to improve the moisturizing effect, cuticle lift resistance, as well as contributing to an increase in tensile strength and waving efficiency. Amino acids are also widely used as essential components in hare care product. Therefore, it is important to investigate how amino acids, peptides, and hydrolyzed proteins diffuse into keratin fibers and how the structural stability of keratin fibers is influenced. However, studies on the penetration of hydrolyzed proteins and amino acids into keratin fibers are still lacking comprehensiveness. Especially, it is difficult to detect whether amino acids penetrated into keratin fibers since amino acids are an essential element of keratin fibers.

The advantage of Raman spectroscopy for studying keratin fibers is that it is nondestructive, requires no sample extraction or purification, and provides information about –SS– groups through reduction and oxidation, which is impossible to record using infrared spectroscopy, since bands can be assigned to S–S and C–S vibrations of cystine. Also, structural information is provided by amide I and amide III vibrations, and the skeletal C–C stretch (α), which is only weakly active in the infrared absorption spectrum of keratin fibers. In addition, aromatic structural information is provided by the amino acids such as tryptophan, tyrosine, and phenylalanine (Phe).

Since the work of Frushour and Koenig has provided assignments for the side and main chain vibrations in wool keratin [7], Raman spectroscopy has been used in structural studies of keratin fibers related to wool finishing [8-15] and hair science [16-31]. We have developed a novel method using Raman spectroscopy [17-29] for directly characterizing the structure of cross-sections at various depths of keratin fibers without isolating the cuticle and cortical cells. When directly characterizing the cuticle and cortex structure of a single keratin fiber, the analytical technique using a Raman microscope is effective since it can be measured at a spot diameter of 1 μm. Most importantly, by using this method, information about the fibrous (keratin) and amorphous (keratin associated protein) structures existing in the cortex region can be obtained. Using this analytical technique, we have been successful in recording the Raman spectra of virgin black human hair, which had been impossible due to high melanin granule content [24, 25, 28]. Also, we have achieved analysis of the heterogeneous reaction (the disconnection of –SS– groups) between the reducing agents (thioglycolic acid and L-cysteine) and virgin white human hair by microspectrophotometry and Raman spectroscopy [18, 21, 25, 27].

In this chapter, the author has described a new method for not only observing the penetration of amino acids and peptides into hair keratin fibers using optical microscopy and dyeing technique, but also directly analyzing the structure of cross-sections at various depths of white human hair treated with Phe using Raman spectroscopy, in order to investigate the penetration of amino acids and peptides into hair keratin fibers.

EXPERIMENTAL

Materials

Virgin black and white Chinese hair bundles (average fiber diameter: 74 μm) which do not perform chemical treatment to be used as keratin fibers were purchased from Beaulax Co. (Tokyo, Japan).

The bleaching cream consists of three components and becomes 5.9 wt % hydrogen peroxide concentration and pH 10.3 when the three components are mixed. Also, other active ingredients, in the bleaching cream, which aid in bleaching are potassium persulfate, ammonium persulfate and sodium persulfate.

The amino acids; L-alanine (Ala), L-valine (Val), L-leucine (Leu), L-methionine (Met), L-phenylalanine (Phe), L-tryptophan (Trp), L-proline (Pro), glycine (Gly), L-cysteine (Cys), L-lysine (Lys), L-aspartic acid (Asp), the peptides; glycylglycine (GlyGly), glycylglycylglycine (GlyGlyGly), and finally methylene blue used as a basic dye and sodium laurylsulfate (SLS) used as a washing agent were purchased from Wako Pure Chemical Industries Ltd. (Osaka, Japan). Also, the peptide; D, L-alanylglycine (AlaGly) was purchased from Tokyo Chemical Industry Co., Ltd. (Tokyo, Japan).

Preparation of Human Hair Treated with Amino Acids and Peptides

Virgin white Chinese hair bundles were immersed in a solution of 0.5 wt % sodium laurylsulfate (SLS) at a ratio of hair to solution of 1: 60. The hair bundles were soaked for 60 min at 50°C. Next, the hair bundles were washed in distilled water and then dried in air (SLS treatment).

As a control, two hair bundles were treated with a phosphate buffer (PB) according to the following procedure. The hair bundles were immersed in a solution of 0.5 M phosphate buffer (pH 7.0) at a ratio of hair: solution = 1: 250. The hair samples were soaked at 50°C for 1 and 16 h. After washing in distilled water for 1 min, the hair samples were then dried at room temperature (PB treatment procedure).

In addition, twenty three human hair bundles were immersed in a solution of 1.0 wt % amino acids and peptides/ 0.5 M phosphate buffer (pH 7.0) at a ratio of hair: solution = 1: 250. The hair samples were soaked at 50°C for 1 and 16 h. After washing in distilled water for 1 min, all samples were dried at room temperature.

Preparation of Human Hair Treated with Phe for Analysis Using Raman Spectroscopy

The SLS treatment procedure was performed on another virgin white Chinese hair bundle.

Sample 1 (Control)

As a control, half of a single virgin white human hair fiber (fiber diameter: 85 μm) was prepared by the following procedures. The hair sample was immersed in a solution of 0.5 M phosphate buffer (pH 7.0) at a ratio of hair: solution = 1: 250. The hair sample was soaked at 50°C for 16 h. After washing in distilled water for 1 min, the hair sample was dried at room temperature (PB treatment procedure).

Sample 2 (Phe)

The other half of the above single human hair fiber was prepared by the following procedures. The hair sample was immersed in a solution of 1 % Phe /0.5 M phosphate buffer (pH 7.0) at a ratio of hair: solution = 1: 250. The hair sample was soaked at 50°C for 16 h. After washing in distilled water for 1 min, the hair sample was dried at room temperature.

Indirect Evaluation of the Penetration of Amino Acids and Peptides into Human Hair

Virgin white human hair fibers treated with amino acids and peptides as described in the previous section were embedded in a resin (Tissue-Tek O.C.T.4583 Compound) and frozen. The frozen blocks were microtomed on a Leica CM1800 (Leica Instruments GmbH, Heidelberger, Germany) to 10 μm thickness, and mounted on a slide glass. Next, the amino acid, peptide, and hydrolyzed protein penetrated parts of the cross sectional samples were dyed with a solution of 0.001 wt % methylene blue using a syringe at room temperature. Finally, the penetration of the amino acids and peptides into the cross-sectional samples was indirectly observed by optical microscopy.

Raman Spectra

All Raman spectra were recorded on a Ramanor T-64000 Raman microscope system (Jobin Yvon, Longjumeau, France), which is comprised of an optical microscope adapted to a single grating spectrograph and a charge coupled device (CCD) array detector (Jobin Yvon, Prism, 1024 × 256 pixel). The laser excitation was provided by an argon ion laser operating at 50 mW of 514.5 nm output. The laser beam on the sample was focused to a spot diameter of 1 μm using a 100× microscope objective. Spectra were recorded by scanning the 200–2000 cm^{-1} region with a total acquisition time of 1000 seconds. One scan with a 1000-second laser exposure was taken in order to obtain a good signal/noise (S/N) ratio. A spectra resolution of 2.3 cm^{-1} was used. By collecting three spectra from the samples, and taking an average of these, it was possible to ensure no sample degradation occurred, and that the spectrum obtained was quite reproducible.

Normalization of Raman spectra of keratin fibers was carried out based on the C–H band at 1448 cm^{-1}, in which the peak area is large and is not influenced by the reduction treatment [18, 19, 21-29]. The phenylalanine (Phe) content of the hair samples was compared by estimating the ratio of the peak area of the Phe peak (calculated from the peak to a baseline which was drawn between 986 and 1020 cm^{-1}) divided by the peak area of the C–H band (calculated from the peak to a baseline which was drawn between 1375 and 1500 cm^{-1}). The mean and standard deviation of phe content in the cuticle and cortex region of hair samples were calculated from the respective contents measured at the five analysis points (Depths of 1, 3, 5, 10, and 20 μm).

Finally, hair samples were embedded in an epoxy resin (Refine Tec Ltd., Yokohama, Japan), and the cured blocks were microtomed on a microtome HM360 (Microm international GmbH, Walldorf, Germany) to 1 μm thickness, and mounted on a slide glass. Cross-sectional samples were produced using virgin white human hair, and sections of the hair at varying depths (1, 3, 5, 10, and 20 μm) from the surface (spot diameter: 1 μm) were measured with a Raman microscope.

RESULTS AND DISCUSSION

Amino Acid Analysis of Human Hair

Bleaching treatments for hair keratin fibers are widely used in the cosmetic industry to lighten the color of human hair, but they cause significant damage. The changes in the chemical properties of human hair by performing bleaching treatments have been extensively studied. It has been found that there is a decrease in the 1/2-cystine content [1, 32-36], an increase in the cysteic acid content [1, 32-36], a decrease in the methionine and tyrosine [1, 32, 33], and an elution of proteins [37], when performing bleaching treatments. Especially, the oxidative cleavage of the -SS- groups that occurs during the chemical bleaching of human hair by current bleaching products is predominately an S–S fission process where the –SS– groups are finally converted to cysteic acid [1, 33]. The author has found that the –SS– content existing from the cuticle region to the center of the cortex region of the virgin black human hair decreased remarkably, while the cysteic acid content increased significantly compared with that of the virgin white human hair by performing the excessive bleaching treatment [22].

In this section, we performed the amino acid analysis on both the virgin and excessively bleached samples of black and white human hair. The hair sample was treated with a bleaching cream (Gatsby Ex Hi-Bleach, Mandom Corp., Osaka, Japan) at 25°C for 30 min at a ratio of hair to solution of 1: 2 and then washed in distilled water for 1 min were prepared. The same procedure was repeated 5 times (bleaching treatment). Finally, the hair sample treated with the bleaching cream was washed in distilled water for 1 min, and then dried at room temperature.

The amino acid compositions of virgin and excessively bleached human hair samples (black and white human hair) are shown in Table 1. The 1/2-cystine content of virgin black human hair (Sample 3) was slightly higher compared with that of virgin white human hair (Sample 1), while the cysteic acid content of Sample 3 was slightly lower compared with that of Sample 1. This result was in agreement with Hollfelder et al.'s opinion [38]. Also, the 1/2-cystine contents of both black and white virgin human hair (Samples 1, and 3) decreased, while the cysteic acid content increased by performing the bleaching treatment. Moreover, the decrease in 1/2-cystine content and the increase in cysteic acid content of the virgin black human hair resulting from the bleaching treatment was larger compared with those of the virgin white human hair resulting from the bleaching treatment. Furthermore, methionine, lysin, and histidine contents of virgin black human hair decreased, while methionine, lysin, and histidine contents of the bleached virgin white human hair did not change by performing the bleaching treatment. This suggests that the not only –SS– groups but also methionine, lysin, and histidine existing in virgin black human hair decomposed compared with those of virgin white human hair by performing the bleaching treatment.

Table 1. Amino Acid Composition of Virgin and Excessively Bleached Human Hair (Black and White Human Hair)

Amino Acid	White Human Hair (mol %)		Black Human Hair (mol %)	
	Virgin	Bleached	Virgin	Bleached
Alanine	4.78	4.70	4.69	4.87
Arginine	6.15	6.34	6.30	6.57
Aspartic Acid	5.75	5.66	5.76	6.10
Cysteic Acid	0.78	4.26	0.32	9.22
1/2-Cystine	14.78	12.92	16.77	7.24
Glycine	6.10	5.89	5.91	5.70
Glutamic Acid	12.86	12.25	12.09	12.81
Histidine	0.88	0.85	0.89	0.75
Isoleucine	2.94	2.92	2.98	3.05
Leucine	6.72	6.65	6.65	7.09
Lysine	2.69	2.64	2.71	2.36
Methionine	0.44	0.42	0.54	0.32
Phenylalanine	1.68	1.66	1.71	1.66
Proline	7.82	7.66	7.47	7.66
Serine	11.39	10.50	10.28	10.25
Threonine	7.26	7.12	7.06	7.17
Tyrosine	1.05	1.75	2.11	1.25
Valine	5.86	5.82	5.75	5.94

Penetration of Amino Acids and Peptides into Virgin Human Hair

Amino acids and peptides have anionic charges above the isoelectric point due to a carboxyl group in the molecule. So, the penetration of amino acids and peptides into virgin human hair can be theoretically observed by dyeing amino acid, and peptide penetrated parts with a basic dye (methylene blue etc.).

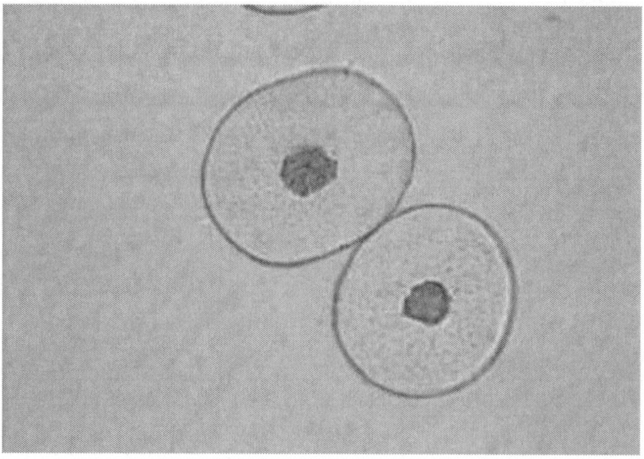

Figure 3. Photomicrograph of the virgin white human hair only treated with 0.5 M phosphate buffer at 50°C and pH 7.0 for 1 h (PB treatment procedure), then cross-sectioned and finally dyed with methylene blue.

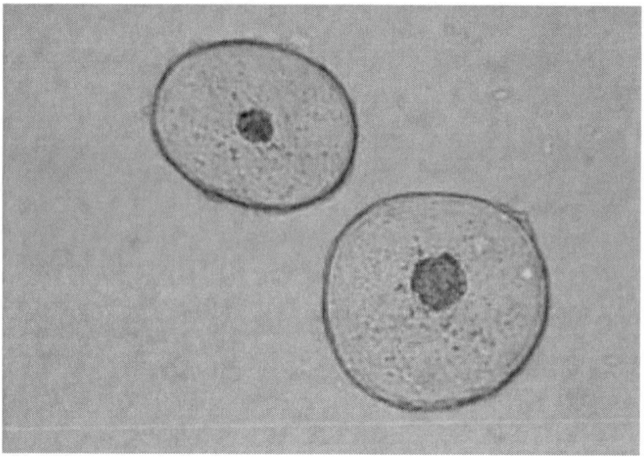

Figure 4. Photomicrograph of the virgin white human hair treated with 1.0 wt % L-phenylalanine (Phe) at 50°C and pH 7.0 for 1 h, then cross-sectioned and finally dyed with methylene blue.

Here, we prepared cross-sectional samples of virgin human hair treated with amino acids and peptides. Next, the penetration of the amino acids and peptides of the cross-sectional samples dyed with methylene blue was indirectly observed by optical microscopy. The photomicrograph of the virgin white human hair only treated with 0.5 M phosphate buffer at 50°C and pH 7.0 for 1 h, then cross-sectioned and finally dyed with methylene blue is shown in Figure 3. The cuticle and the cortex of the untreated white human hair sample did not adsorb the methylene blue. However, the medulla, which exists in the center of hair fibers, adsorbed the methylene blue, since the medulla is rich in glutamic acid and consists of porous proteins. The photomicrographs of the virgin white human hair treated with 1.0 wt % L-phenylalanine (Phe) and glycine (Gly) at 50°C and pH 7.0 for 1 h, then cross-sectioned and finally dyed with methylene blue, are shown in Figures 4 and 5, respectively. The virgin white human hair treated with Phe at 50°C and pH 7.0 for 1 h, adsorbed the methylene blue through the cuticle and partially into the cortex (Figure 4). On the other hand, the virgin white human hair treated with Gly at 50°C and pH 7.0 for 1 h, did not adsorb the methylene blue into the cuticle and cortex (Figure 5). This suggests that the penetration of amino acids into the human hair was influenced by the hydrophobic property of amino acids.

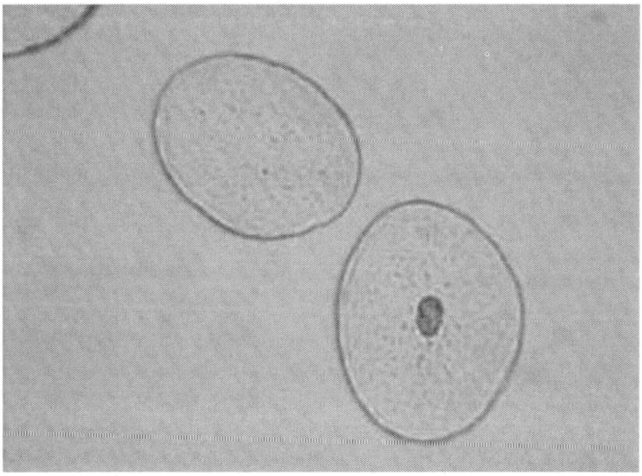

Figure 5. Photomicrograph of the virgin white human hair treated with 1.0 wt % glycine (Gly) at 50°C and pH 7.0 for 1 h, then cross-sectioned and finally dyed with methylene blue.

The penetration of amino acids and peptides into virgin human hair is shown in Table 2. In the case of performing the amino acid treatment at 50°C and pH 7.0 for 1 h, the hydrophobic amino acids, especially Phe and L-tryptophan (Trp) which have a aromatic ring, most deepely penetrated into the virgin white human hair. Also, the penetration of the hydrophobic amino acids increased by increasing the methylene groups in the side chain of the hydrophobic amino acids (Ala<Val <Leu). On the other hand, the hydrophilic amino acids (Asp, Cys, Gly, Lys) and peptides (GlyGly, AlaGly, GlyGlyGly) did not penetrate at all. In the case of performing the amino acid treatment at

50°C and pH 7.0 for 16 h, the amino acids (Gly, Ala, Met) and peptides (GlyGly, AlaGly) penetration except for Pro, did not change at all.

Table 2. Penetration of Amino Acids and Peptides into Virgin Human Hair

Agents	50°C, 1 h		50°C, 16 h	
	Cuticle	Cortex	Cuticle	Cortex
Control	–	–	–	–
Ala	+	–	+	–
Val	ND[a]	ND	+	–
Leu	+	± ~ ++	+	± ~ ++
Met	+	–	+	–
Phe	+	± ~ ++	+	± ~ ++
Trp	+	± ~ ++	+	± ~ ++
Pro	+	–	+	± ~ ++
Gly	–	–	+	–
Cys	+	–	ND	ND
Lys	–	–	ND	ND
Asp	+	–	ND	ND
GlyGly	–	–	–	–
AlaGly	– ~ ±	–	+	–
GlyGlyGly	–	–	ND	ND

–: Did not penetrate; ±: Trace; +: Penetrated; ++; Deeply penetrated into cortex region.
[a] Not done.

Raman Spectra of Hair Keratin Fibers

Measurement by Raman spectroscopy becomes a beneficial means of investigating the structural changes of cross-sections at various depths of human hair due to it being able to obtain information on the secondary structure of proteins and disulfide (–SS–) groups in keratin fibers.

The bands of particular interest lie in the wave number range of 500-1800 cm^{-1}. These are vibrations assigned to the S–S and C–S bonds of cystine; the amino acids tryptophan, tyrosine, and phenylalanine; the amide I and amide III vibrations; and the C–C skeletal stretching vibration of the α-helix. The Raman spectra of the virgin white human hair fiber at depths of 1, 5, 10, and 30 μm are shown in Figure 6. The depth of 1 μm from fiber surface corresponds to the cuticle region and the depth of 5 ~ 30 μm from fiber surface corresponds to the cortex region. It is shown that the band shapes, as well as peak maximum frequencies, of the cuticle region were significantly different from those of the cortex region of the human hair fibers. The frequencies and tentative assignments of the virgin white human hair fiber at depths of 1 μm and 5 μm (corresponding to the cuticle and cortex) from the fiber surface compared with that of wool are shown in Table 3.

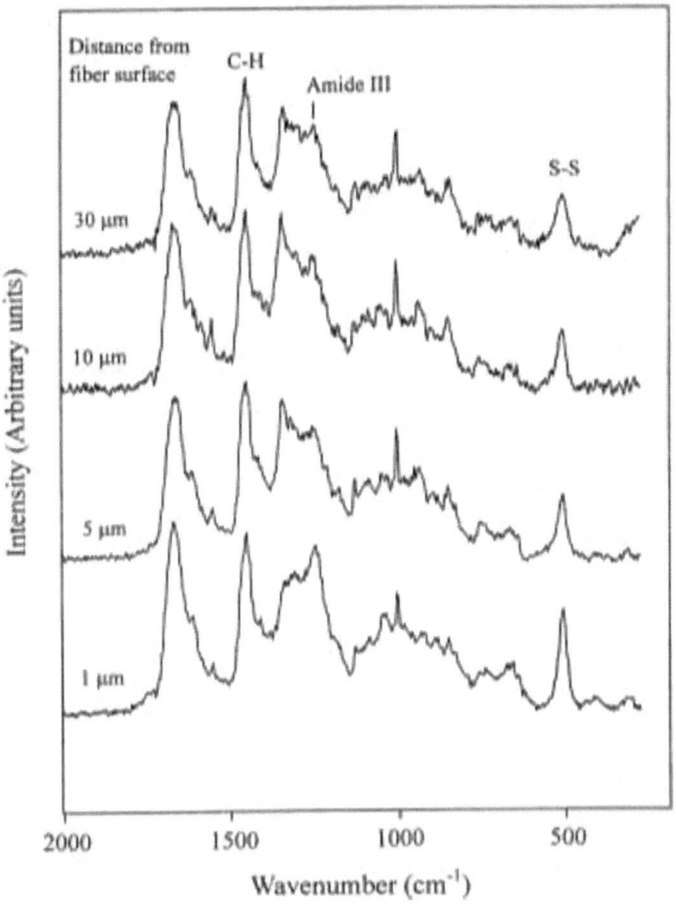

Figure 6. Raman spectra of the virgin white human hair at depths of 1, 5, 10, and 30 μm.

Table 3. Frequencies and Tentative Assignments of Untreated Human Hair (Cuticle and Cortex Region) Compared with that of Wool

Human Hair		Wool		Assignment
Cuticle [a] (cm^{-1})	Cortex [b] (cm^{-1})	Ref. 11 (cm^{-1})	Ref. 7 (cm^{-1})	
1671	1666	1658	1653	Amide I
1613	1614	1615	1614	Tyr and Trp
1553	1552	1558	1553	Trp
1448	1446	1450	1448	CH_2 bending mode
ND [c]	1336	1340	1338	CH_2 bend, Trp
ND	1315	1316	1318	Cα-H bend
1245	1243	1245	1244	Amide III (unordered)
ND	1210	1209	1207	Tyr and Phe
ND	1174	1180	1176	Tyr
1123	1123	1126	1126	C-N stretch
1040	1040	–	–	Sulfonate S-O stretch
ND	1030	1034	1031	Phe
1001	1001	1006	1002	Phe
959	ND	959	952	CH_2 rock
ND	935	935	934	Skeletal C-C stretch (α)
884	880	883	881	Trp
851	851	852	851	Tyr
ND	750	752	752	Trp
664	664	665	661	Cys C-S stretch
642	642	644	642	Tyr
505	507	512	512	Cys S-S stretch g-g

[a] Depth of 1 μm from hair surface.

[b] Depth of 5 μm from hair surface.

[c] Not detected.

The amide I peak maxima for the cortex region (depth of 5 μm from fiber surface) was found to shift at 1666 cm^{-1}. Also, the skeletal C–C stretch (α) band at 935 cm^{-1}, assigned to the α-helical backbone, was observed in the cortex region only. This is in agreement with the findings by Fraser et al., in which the microfibril that exists in the cortex region is mainly composed of α-helical protein [4, 39]. On the other hand, the amide I peak maxima for the cuticle region (depth of 1 μm from fiber surface) was found to be at 1671 cm^{-1}, assigned to the β-sheet and/or random coil forms, but the skeletal C–C stretch (α) band, assigned to the α-helical backbone did not appear. Also, the amide III (unordered) band intensity, assigned to the random coil form, at 1245 cm^{-1} for the cuticle region was higher than that for the cortex region. These results indicate that the α-helix form does not exist in the hair cuticle, and that the cuticle has a more amorphous structure. Furthermore, the S–O band, assigned to cysteic acid, was observed at 1040 cm^{-1} for the cuticle region whereas the S–O band for the cortex region was very weak.

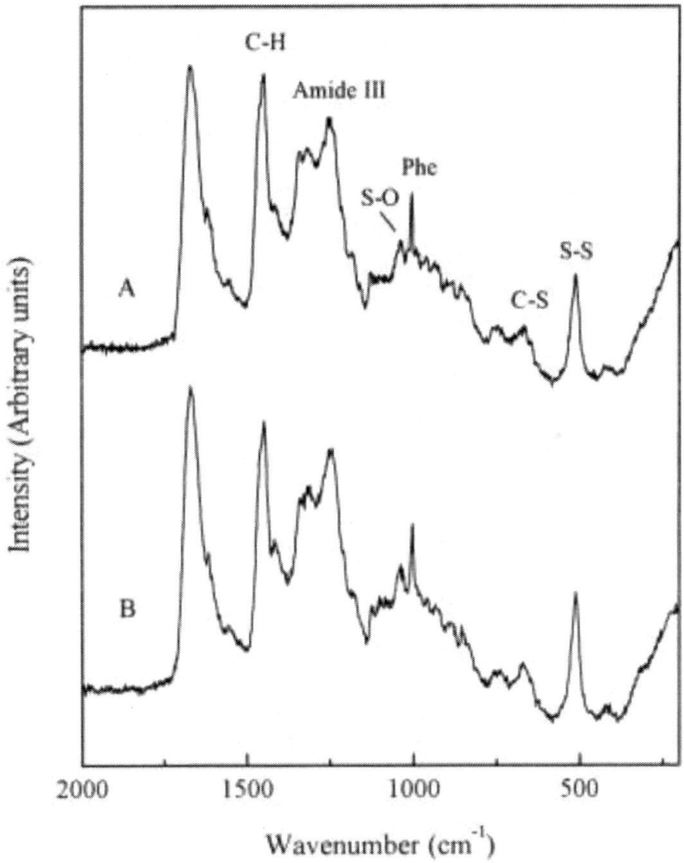

Figure 7. Cuticle Raman spectra (depth of 1 μm from the fiber surface) of the virgin white human hair (Sample 1) and the virgin white human hair treated with Phe (Sample 2): (A) Sample 1; (B) Sample 2.

Raman Spectra of Hair Keratin Fibers Treated with Phe

Measurement by Raman spectroscopy becomes a beneficial means of investigating the penetration of hydrolyzed proteins and aromatic amino acids into human hair due to it being able to obtain information on the amino acids tryptophan, tyrosine, and phenylalanine, the S–S and C–S bonds of cystine, and the secondary structures, such as random coil, β-sheet, and α-helix in keratin fibers.

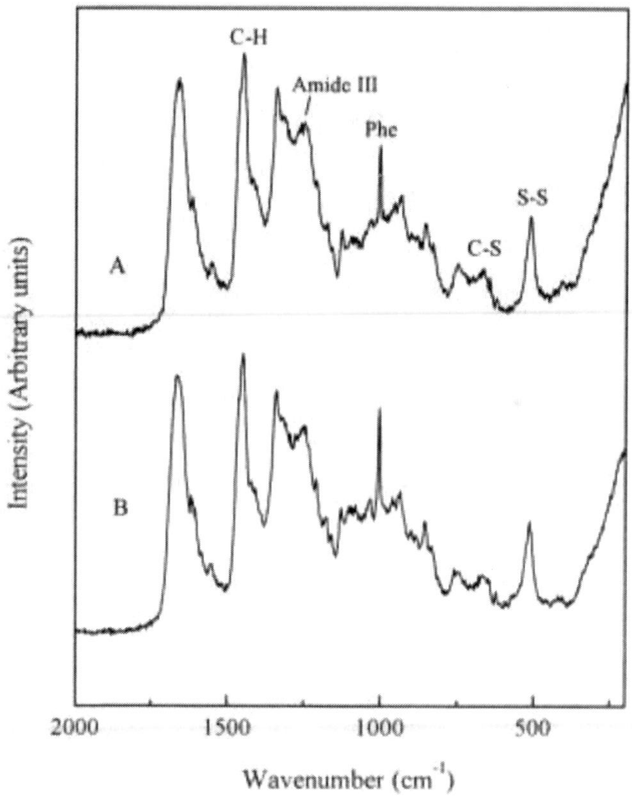

Figure 8. Representative cortex Raman spectra (depth of 5 μm from the fiber surface) of the virgin white human hair (Sample 1) and the virgin white human hair treated with Phe (Sample 2) (A) Sample 1; (B) Sample 2.

In this section, the white human hair samples treated with a solution of 1.0 wt % Phe (pH = 7.0) at 50°C for 16 h were analyzed at various depths using a Raman microscope. The cuticle Raman spectra (depth of 1 μm from the fiber surface) of the virgin white human hair (Sample 1) and the virgin white human hair treated with Phe (Sample 2) is shown in Figure 7. As is shown, the band shape, as well as peak maximum frequency existing in the cuticle region of the virgin white human hair were almost unchanged by performing the Phe treatment.

The representative cortex Raman spectra (depth of 5 μm from the fiber surface) of the virgin white human hair (Sample 1) and the virgin white human hair treated with Phe (Sample 2) is shown in Figure 8. As is shown, the band shape, as well as peak maximum frequency existing in the cortex region of the virgin white human hair were almost unchanged by performing the Phe treatment. Also, it is shown that the band shape, as well as peak maximum frequency, of the cuticle region of the human hair fiber were significantly different from those of the cortex region of the human hair fiber.

Next, to investigate the influence of the Phe treatment on the cuticle and cortex of human hair, the Phe content at various depths of the hair fibers untreated and treated with Phe was compared by Raman spectroscopic analysis. Normalization of Raman spectra of keratin fibers is often carried out based on the C–H band at 1450 cm^{-1} [18, 19, 21-29], amide I band at 1666 cm^{-1} [11], and the Phenylalanine (Phe) peak at 1001 cm^{-1} [15, 17, 20]. Here, we chose the C–H band, which is assigned at 1448 cm^{-1}, for normalization because the peak area is large and is not influenced by the Phe treatment.

The Phe content in hair samples (Samples 1 and 2) at depths of 1, 3, 5, 10, and 20 μm from the fiber surface are shown in Table 4. The Phe content (0.066) of virgin white human hair (Sample 1: Control) for the cortex region (depth of 3 μm from the fiber surface) increased compared with that (0.046) of the cuticle region (depth of 1 μm from the fiber surface) ($P < 0.001$). The ratio of Phe content in the cuticle and cortex region (Phe content of cuticle/ Phe content of cortex = 0.046/ 0.066) was 0.70. This ratio is in excellent agreement with that obtained from the amino acid analysis (11.5/ 16.5 = 0. 70), namely the Phe content of the cuticle and cortex were 11.5 mol/ 1000 mol total amino acids and 16.5 mol/ 1000 mol total amino acids, respectively [32]. This result indicates that the Phe content of the cuticle and cortex can be reliably estimated using Raman spectroscopic analysis.

The Phe content of Sample 2 (the Phe treatment) at the distance of 1, 3, 5, 10, and 20 μm from the fiber surface were clearly higher than that of Sample 1 (Control), indicating that the Phe penetrated into virgin white human hair by performing the Phe treatment.

Table 4. Phe Content in Hair Samples (Samples 1 and 2) at Depths of 1, 3, 5, 10, and 20 μm from the Fiber Surface

Sample	Distance from fiber surface (μm)				
	1	3	5	10	20
1 (Control)	0.05	0.07	0.06	0.07	0.06
	0.04	0.06	0.06	0.06	0.07
	0.05	0.07	0.06	0.07	0.07
	0.05	0.06	0.06	0.06	0.06
	0.04	0.07	–	–	–
Mean ± SD	*0.046 ± 0.005*	*0.066 ± 0.005*	*0.060 ± 0.000*	*0.065 ± 0.005*	*0.065 ± 0.005*
2 (Phe)	0.06	0.09	0.09	0.08	0.08
	0.06	0.09	0.08	0.08	0.08
	0.06	0.09	0.08	0.08	0.09
	0.06	0.09	0.07	–	0.08
	0.06	0.09	–	–	–
Mean ± SD	*0.060 ± 0.000*	*0.09 ± 0.000*	*0.080 ± 0.007*	*0.080 ± 0.000*	*0.083 ± 0.004*

From these experiments, Phe was found to have more deeply penetrated the virgin white human hair samples by performing the Phe treatment at pH 7.0 for 16 h.

CONCLUSION

We have developed a new method for investigating the penetration of amino acids and peptides into hair keratin fibers, which until now had not been successfully achieved. In addition, it was revealed that the penetration of L-phenylalanine (Phe) into virgin human hair could be directly evaluated by measuring the Phe content using Raman spectroscopy. As a result, it has been found that hydrophobic amino acids, such as phe, L-tryptophan (Trp), and L-leucine (Leu), deeply penetrated into the virgin human hair, while the hydrophilic amino acids and peptides did not penetrate at all. Considering that the apolar amino acids into two strand coiled-coil ropes acts as a hydrophobic interaction, the penetration of hydrophobic amino acids into human hair may leads to an increase of tensile strength resulting in healthier hair.

REFERENCES

[1] Robbins, CR. *Chemical and Physical Behavior of Human Hair.* 2rd ed. New York/Berlin/Heidelberg: Springer-Verlag; 1988.

[2] Lindley, H. The Chemical Composition and Structure of Wool. In: Asquith RS, editor. *Chemistry of Natural Protein Fibers.* New York: Plenum Press; 1977; pp 147-191.

[3] Arai, K. The Chemistry of Wool and Its Structure and Properties. *Sen-i Gakkaishi 1989*, 45 (12), 512-516.

[4] Fraser, RDB; Gillespie, JM; MacRae, TP; Marshall, RC. *Schematic diagram of a fine wool fiber* (CSIRO Division of Protein Chemistry); In the paper referenced by Marshall, RC; Gillespie, JM; McGuirk, GJ; Marler, JW; Reis, PJ; Rogen, IM; Whiteley, KJ. Proc 7th Int Wool Text Res Conf, Tokyo 1985, II, pp 36-44.

[5] Gillespie, JM. The proteins of Hair and Other Hard α-Keratins. In: Goldman RD, Steinert PM editors. *Cellular and Molecular Biology of Intermediate Filaments.* New York/London: Plenum Press; 1990; pp 95-128.

[6] Feughelman, M. Natural Protein Fibers. *J Appl Polym Sci* 2002, 83, 489-507.

[7] Frushour, BG; Koenig, JL. Raman spectroscopy of proteins. In: Clark RJH, Hester RE editors. *Advances in Infrared and Raman spectroscopy:* Vol. 1. London: Heyden; 1975; pp 35-97.

[8] Hsu, SL; Moore, WH; Krimm, S. Vibrational Spectrum of the Unordered Polypeptide Chain: A Raman Study of Feather Keratin. *Biopolymers* 1976, 15, 1513-1528.

[9] Shishoo, R; Lundell, M. Investigation of Structural Changes in Wool Fibers Due to Annealing. *J Polym Sci Polym Chem Ed* 1976, 14, 2535-2544.

[10] Hogg, LJ; Edwards, HGM; Farwell, DW; Peters, AT. FT Raman Spectroscopic Studies of Wool. *J Soc Dyers Colour* 1994, 110, 196-199.

[11] Carter, EA; Fredericks, PM; Church, JS; Denning, RJ. FT-Raman Spectroscopy of Wool – I. *Preliminary Studies. Spectrochem Acta* 1994, 50A, 1927-1936.

[12] Church, JS; Corino, GL; Woodhead, AL. The Analisis of Merino Wool Cuticle and Cortical Cells by Fourier Transform Raman Spectroscopy. *Biopolymers,* 1997, 42, 7-17.

[13] Jones, DC; Carr, CM; Cooke, WD; Lewis, DM. Investigating the Photo-Oxidation of Wool Using FT-Raman and FT-IR Spectroscopies. *Textile Res J* 1998, 68 (10), 739-748.

[14] Rintoul, L; Carter, EA; Stewart, SD; Fredericks, PM. Keratin Orientation in Wool and Feathers by Polarized Raman Spectroscopy. *Biopolymers* 2000, 57, 19-28.

[15] Kuzuhara, A; Hori, T. Reducing Wrinkle Formation in Wool with 2-Iminothiolane Hydrochloride. *Textile Res J* 2002, 72, 285-289.

[16] Pande, CM. FT-Raman Spectroscopy – Applications in hair research. *J Soc Cosmetic Chem 1994*, 45, 257-268.

[17] Kuzuhara, A. Chemical Modification of Keratin Fibers Using 2-Iminothiolane Hydrochloride. *J Appl Polym Sci* 2003, 90, 3646-3651.

[18] Kuzuhara, A; Hori, T. Reduction Mechanism of Tioglycolic Acid on Keratin Fibers Using Microspectrophotometry and FT-Raman Spectroscopy. *Polymer* 2003, 44, 7963-7980.

[19] Kuzuhara, A. Analysis of Structural Change in Keratin Fibers Resulting from Chemical Treatments Using Raman Spectroscopy. *Biopolymers* 2005, 77, 335-344.

[20] Kuzuhara, A. Protein Structural Changes in Keratin Fibers Induced by Chemical Modification Using 2-Iminothiolane Hydrochloride: A Raman Spectroscopic Investigation. *Biopolymers* 2005, 79, 173-184.

[21] Kuzuhara, A; Hori, T. Reduction Mechanism of L-Cysteine on Keratin Fibers Using Microspectrophotometry and Raman Spectroscopy. *Biopolymers* 2005 79, 324-334.

[22] Kuzuhara, A. Analysis of Structural Changes in Bleached Keratin Fibers (Black and White Human Hair) Using Raman Spectroscopy. *Biopolymers* 2006, 81, 506-514.

[23] Kuzuhara, A. Analysis of Structural Changes in Permanent Waved Human Hair Using Raman Spectroscopy. *Biopolymers* 2007, 85, 274-283.

[24] Kuzuhara, A.; Fujiwara, N.; Hori, T. Analysis of Internal Structure Changes in Black Human Hair Keratin Fibers with Aging Using Raman Spectroscopy. *Biopolymers* 2007, 87, 134-140.

[25] Kuzuhara, A. A *New Method of Internal Structural Analysis of Keratin Fibers Using Raman Spectroscopy*. In: Nemeth TS editors. Biopolymer Research Trends: New York: Nova Science Publishers Inc.; 2007; pp 49-86.

[26] Kuzuhara, A. Raman Spectroscopic Analysis of L-Phenylalanine and Hydrolyzed Eggwhite Protein Penetration into Keratin Fibers. *J Appl Polym Sci* 2011, 122, 2680-2689.

[27] Kuzuhara, A. Analysis of Heterogeneous Reaction between Reducing Agents and Keratin Fibers Using Raman Spectroscopy and Microspectrophotometry. *J Mol Struct* 2013, 1037, 85-92.

[28] Kuzuhara, A. Analysis of Internal Structure Changes in Black Human Hair Keratin Fibers Resulting from Bleaching Treatments Using Raman Spectroscopy. *J Mol Struct* 2013, 1047, 186-193.

[29] Kuzuhara, A. Internal Structure Changes in Bleached Black Human Hair Resulting from Chemical Treatments: A Raman Spectroscopic investigation. *J Mol Struct* 2014, 1076, 373-381.

[30] Schlucker, S; Liang, C; Strehle, KR; DiGiovanna, JJ; Kraemer, KH; Levin, IW. Conformational Differences in Protein Disulfide Linkages between Normal Hair and Hair from Subjects with Trichothiodystrophy: A Quantitative Analysis by Raman Microspectroscopy. *Biopolymers,* 2006, 82, 615-622.

[31] Paquin, R.; Colomban, P. Nanomechanics of Single Keratin Fibers: A Raman Study of the α-Helix → β-Sheet Transition and the Effect of Water. *J Raman Spectrosc* 2007, 38, 504-514.

[32] Chao, J; Newson, AE; Wainwright, IM; Mathews, RA. Comparison of the Effects of Some Reactive Chemicals on the Proteins of Whole Hair, Cuticle and Cortex. *J Soc Cosmetic Chem* 1979, 30, 401-413.

[33] Robbins, CR. *Chemical and Physical Behavior of Human Hair.* 4th ed. New York/Berlin/Heidelberg: Springer; 2001.

[34] Robbins, CR; Kelly, C. Amino Acid Analysis of Cosmetically Altered Hair. *J Soc Cosmetic Chem* 1969, 20, 555-564.

[35] Wolfram, LJ; Hall, K; Hui I. The Mechanism of Hair Bleaching. *J Soc Cosmetic Chem* 1970, 21, 875-900.

[36] Erlemann, GA; Beyer, H. Influence of Hydrogen-Peroxide to the Chemical Structure of Human Hair. *J Soc Cosmetic Chem* 1971, 22, 795-807.

[37] Sandhu, SS; Robbins, CR. A Simple and Sensitive Technique, Based on Protein Loss Measurement, to Assess Surface Damage to Human Hair. *J Soc Cosmetic Chem* 1993, 44, 163-175.

[38] Hollfelder, B; Blankenburg, G; Wolfman, LJ; Hocker, H. Chemical and Physical Properties of Pigmented and Non-pigmented Hair (Gray Hair). *Int J Cosmet Sci* 1995, 17, 87-89.

[39] Fraser, RDB; MacRae, TP; Rogers, GE. Molecular Organization in Alpha-Keratin. *Nature* 1962, 193, 1052-1055.

INDEX

A

acetone, 57, 58
acid, vii, x, 1, 11, 17, 18, 34, 37, 51, 52, 53, 54, 56, 62, 66, 67, 68, 71, 72, 74, 76, 78, 79, 81, 85
acidity, 54
acrylic acid, 20
acrylonitrile, 20
active site, 18
adolescents, 7, 9
adrenaline, vii, 1, 21
adsorption, 20
adulthood, viii, 8, 29, 30, 32, 34, 44, 47
age, 5, 6, 32, 33, 34, 36, 37, 38, 42, 43, 46
aggression, 8
alanine, 51, 61, 72, 75
alcohols, 50, 53, 64
aldehydes, 50
alternative treatments, 4
amine(s), 13, 14, 15, 21, 50, 61, 62, 63, 66, 67, 69
amino acid(s), vii, ix, 1, 3, 16, 17, 18, 19, 20, 21, 30, 33, 34, 35, 37, 38, 40, 44, 45, 49, 50, 53, 54, 60, 61, 62, 63, 64, 65, 66, 71, 72, 73, 74, 75, 76, 78, 79, 80, 81, 82, 86, 87, 88
amino groups, 17
ammonia, 3, 4, 18, 30, 51, 55
ammonium, 50, 63, 66, 67, 75
amphetamines, 13, 14
analgesic, 18
ANS, 55
anthocyanin, 52, 53, 63, 66, 68, 69
anthropometric characteristics, viii, 29, 30, 31, 32, 33, 38, 42, 45
antibody, 18
antidepressant, viii, 2
antioxidant, 16, 51, 53, 63, 69
aorta, 7
arginine, 54, 72
argon, 77
aromatic compounds, 51, 53, 60, 62
aromatic quality, ix, 49, 50, 53, 60
aromatic ring(s), vii, 1, 18, 21, 81
arthritis, 18
aspartame, vii, 2, 10, 17
aspartate, 11
aspartic acid, 51, 61, 75
asthma, 10
autonomy, 8
autosomal recessive, vii, 2, 30

B

bacteria, 3, 13
bacterium, viii, 2
behavioral problems, 8
benefits, 31, 62
benzyl group, vii, 1

beverages, 61, 62, 69
biochemistry, 9
biological fluids, 2
biological roles, vii, 2
biopsy, 4
biosynthesis, 51, 52, 66, 67, 69
biotic, 51
births, 2
bleaching, 75, 78
blood, vii, 2, 3, 5, 6, 7, 8, 9, 13, 30, 31, 33, 34, 36, 38, 42, 47
blood pressure, 13
blood vessels, 13
blood-brain barrier, viii, 2
body mass index (BMI), viii, 29, 30, 32, 32, 33, 37, 38, 39, 41, 42, 43, 47
bonding, 16
bonds, 82, 86
brain, vii, 2, 6, 9, 10, 11, 13, 15
brain damage, vii, 2
brain functioning, 11
branching, 51
breast milk, vii, 1

C

cancer, 3, 51
carbohydrates, 17, 53, 63
carboxyl, 17, 79
carboxylic acid(s), 18, 50
cardiac output, 15
casein, 2
catabolism, 50
catecholamines, vii, 1, 13, 21
Caucasian population, 4
causal relationship, 2
C-C, 84
cell culture, 63, 67
cellulose, 19
charge coupled device, 77
chemical, 13, 17, 18, 52, 54, 73, 75, 78
chemical properties, 78
chitosan, 19
chromatography, 19, 21
chromosome, 3
chronic diseases, 51
cleavage, 78
clusters, 59
C-N, 84
coarctation, 7, 9
cocoa, 13
cognitive development, 2, 21, 30
composition, ix, 50, 51, 52, 53, 54, 57, 62, 64, 65, 66
compounds, vii, ix, 1, 3, 12, 15, 17, 18, 21, 49, 50, 52, 53, 54, 57, 58, 59, 60, 61, 62, 63, 64, 67, 68, 69
congenital heart disease, 6, 7, 9
conjugation, 18
consumption, 64
conversion rate, 8
COOH, 17
corpus callosum, 6
correlation, 6, 8, 9, 34, 62
correlation analysis, 34
correlation coefficient, 34
cortex, x, 71, 72, 74, 77, 78, 80, 82, 83, 84, 86, 87
cosmetic, 78
creatine, 9, 44
cultivation, 65
culture, 3, 69
cuticle, x, 71, 72, 73, 74, 77, 78, 80, 83, 84, 86, 87
cysteine, 3, 17, 51, 74, 75
cystine, 3, 72, 74, 78, 82, 86

D

D-amino acids, vii, 2, 17, 18, 20
defects, 6, 9
deficiencies, 63
deficiency, 3, 4, 5, 46, 69
deficit, 31, 44
degradation, 77
depression, 6, 8, 10, 15
depth, 83, 84, 85, 86, 87
derivatives, 13, 16, 17, 18, 21
detection, 18, 45, 63
diabetes, 6

diabetic patients, 6
diet, viii, 2, 3, 4, 5, 7, 9, 10, 13, 21, 29, 30, 31, 33, 34, 36, 37, 38, 39, 40, 41, 42, 43, 44, 45, 47
diffusion, 19, 21
discrimination, 14
discs, 3
diseases, 7, 10, 32, 36
disorder, vii, 2, 5
distilled water, 75, 76, 78
distress, 61
distribution, 4, 15
DNA, 2, 7, 17, 19, 21
DOI, 24
dopamine, vii, 1, 13, 15, 21
dosing, 10
drug addict, 14
drugs, 13, 16, 21
dyeing, ix, 71, 74, 79

E

eczema, 3, 10
electric field, 64, 65, 67
electrophoresis, 5
emission, 17
enantiomers, 19, 20, 21
energy, 44
enology, ix, 50
environmental stress, 53
enzyme(s), vii, 2, 4, 5, 17, 18, 30, 50, 52
epidemiologic, 6
epigenetics, 44
epilepsy, 3
epinephrine, vii, 1
equilibrium, 20
erosion, 18
ester, 10
eukaryotic, 18
evolution, 33, 37, 64, 67
exocytosis, 13
exposure, 6, 77
extraction, 69, 74

F

fasting, 8, 9
fermentation, ix, 49, 50, 52, 54, 57, 59, 60, 61, 62, 63, 64, 65, 66, 67, 68, 69
fertilizers, 65
fiber(s), ix, 16, 71, 72, 73, 74, 75, 76, 77, 78, 81, 82, 83, 86, 87, 88, 89
fine wool, 72, 89
fission, 78
flavonoid pathway, ix, 49
flavonoids, 51
flavonol, 52
flavo(u)r, 50, 57
fluorescence, 16, 17
food, 13, 34, 66
forebrain, 11
formation, 11, 16, 51, 52, 57, 62, 64, 67, 69
formula, 4, 32
freedom, 4
fungi, 13

G

GABA, 11
gait, 3
gastrointestinal tract, 18
gel, 5, 16
gene expression, 63, 67
gene therapy, 4
genes, 3
genetics, 45
genotype, 5
Germany, 76, 77
gestation, 7
gestational age, 47
glucose, 55
glutamate, 11
glutamic acid, 72, 81
glycine, 3, 11, 75, 81
growth, viii, ix, 3, 6, 29, 30, 31, 32, 33, 34, 36, 43, 44, 46, 47, 48, 50, 52
growth impairment, viii, ix, 29, 34, 43, 44
growth rate, viii, 29, 30, 32

H

hair, ix, 71, 72, 73, 74, 75, 76, 77, 78, 79, 80, 81, 82, 83, 84, 85, 86, 87, 88, 90
hair keratin, ix, 71, 74, 78, 88
haplotypes, 4
headache, 10
health, vii, 45, 51, 57, 59, 63
health status, 45
heart rate, 16
height, viii, 29, 30, 32, 33, 34, 36, 37, 38, 39, 41, 42, 44, 47
hemoglobin, 6
histidine, 79
human brain, 2
human health, 51
hydrogen, 16, 20, 75
hydrogen peroxide, 20, 75
hydrolysis, 51
hydrophilic amino acids, ix, 71, 81, 88
hydrophobic, vii, ix, 1, 16, 17, 71, 72, 73, 81, 88
hydroxylation, vii, 2, 8
hyperactivity, 7, 10
hyperphenylalaninaemia, viii, 29, 30, 46
hypoplasia, 6
hypotension, 61

I

in vitro, 4
in vivo, 4, 8, 9
individuals, viii, ix, 10, 29, 30, 33, 34, 37, 43, 44
induction, 13
industry, 78
infants, 3, 6, 7, 9
inflammation, 18
inflammatory bowel disease, 17
infrared spectroscopy, 74
inhibition, 3, 13
inhibitor, 11, 16
initiation, 39, 41
intelligence quotient, 5
intervention, 33, 43
intrauterine growth retardation, 7
introversion, 8
ions, 50
isolation, 6, 8
isoleucine, 54, 61, 69, 72
isomers, 15
isotope, 8, 59, 66

K

keratin, ix, 71, 72, 73, 74, 75, 77, 78, 82, 86, 87, 88
kinetics, 16, 50, 52

L

lack of control, vii, 2
lead, 6, 45
leucine, 72, 75, 88
light, 3, 16, 51, 52, 69
light scattering, 16
liquid chromatography, 11
liver, 4, 20
locomotor, 14
longitudinal study, viii, 29, 30, 31, 36, 43
low-protein diet, viii, 29
lysine, 72, 75

M

magnetic resonance imaging, 5, 6, 10
magnetic resonance spectroscopy, 9
malnutrition, 30, 43
mammalian brain, 13, 15
mammals, vii, 1
mapping, 5
mass spectrometry, 5
measurement(s), 32, 37
meat, 33
median, 6, 34, 35, 40
medical, viii, 2
medicine, 13
medulla, 72, 81

melanin, 74
melatonin, 64
membranes, 19, 20, 21
memory formation, 11
mental retardation, 6, 11
metabolism, vii, viii, 2, 12, 13, 15, 21, 28, 30, 44, 50, 52, 54, 57, 63
metabolites, 51, 52
metabolized, 12, 21
methylene blue, 75, 76, 79, 80, 81
microcephaly, 6, 7
microorganisms, 66
microscope, 74, 77, 86
microtome, 77
molecular mass, 61
molecular weight, 20
molecules, 16, 73
monomers, 67
motivation, 7
MRI, 5, 10
mutagenesis, 18
mutant, 30
mutation(s), 3, 4, 5, 7, 36

N

nervous system, 15
nervousness, 6, 8
neurodegeneration, 13
neurological damage, viii, 29, 30
neurological disease, 15
neurons, 11, 13
neurotransmitter(s), 8, 13, 15
neutral, 30
NH2, 20
niacin, viii, 2
nitrogen, ix, 49, 50, 52, 53, 54, 57, 60, 62, 63, 64, 65, 66, 67, 68
nitrogen compounds, 50, 53, 60, 64, 65, 67, 68
NMR, 16, 18
noradrenaline, vii, 1, 13, 21
norepinephrine, vii, 1
normal development, viii, 29, 30, 33
Norway, 2
nucleotides, 50
nutrient(s), 30, 34, 44, 67
nutrition, viii, 2, 29, 30, 31, 33, 37, 46, 47, 63
nutritional status, 31, 48

O

oligomerization, 5
optical microscopy, ix, 71, 74, 76, 80
organ, 15
ornithine, 3
oxidation, 5, 44, 74

P

PAH, vii, 2, 3, 4, 5, 7, 21, 22, 30, 31, 33, 36, 43, 44, 46
pain, 18
parents, 4, 34
participants, 36
pathogens, 52
pathophysiology, 48
pathways, 36
PCR, 4
peptides, ix, 16, 18, 50, 71, 73, 74, 75, 76, 79, 80, 81, 88
perfusion, 15
permeability, 19
permeation, 20
pH, 17, 54, 75, 76, 80, 81, 86, 88
pharmacology, 15
phase inversion, 19, 20
Phe, vii, ix, 1, 2, 3, 4, 5, 6, 7, 8, 9, 10, 11, 16, 17, 18, 19, 20, 30, 31, 33, 34, 35, 36, 37, 38, 40, 42, 43, 44, 45, 71, 74, 75, 76, 77, 80, 81, 82, 84, 85, 86, 87, 88
phenolic compounds, ix, 49, 51, 52, 57, 62, 68
phenotype(s), 4, 5, 8, 9, 30, 33
phenylalanine hydroxylase, vii, 2, 3, 21, 46
phenylalanine-free formulas, viii, 29
phenylketonuria (PKU), vii, viii, 1, 2, 3, 4, 6, 7, 8, 9, 10, 11, 16, 21, 27, 28, 29, 30,

31, 32, 32, 33, 34, 35, 36, 37, 40, 42, 43, 44, 45, 46, 47, 48
phosphate, 75, 76, 80
phosphocreatine, 9
photomicrographs, 81
physical and mechanical properties, 72
physiology, 15
pigmentation, 3
plant growth, 53
plants, 13, 51, 52, 61
plasma levels, 6
plasma membrane, 13
polymerization, 19
polymer(s), 19, 21
polyphenols, 51, 52, 63
population, viii, 7, 29, 30, 32, 37, 38, 42, 44, 46
positive correlation, 34, 36
positive emotions, 8
potassium, 75
potassium persulfate, 75
pregnancy, 6, 7, 9
prodrugs, 17
proline, ix, 49, 50, 54, 65, 75
protection, 9, 51, 52
protein intake, viii, ix, 29, 30, 31, 34, 40, 43, 44
protein-protein interactions, 4
protein-restricted diets, viii, 29, 30, 31, 43, 44
proteins, vii, 2, 31, 44, 50, 63, 72, 73, 78, 81, 82, 86, 89
psychoactive drug, 13
psychosocial dysfunction, 42
puberty, 33
purification, 74

Q

quality of life, 31

R

race, 16, 19, 20, 32, 36
radical copolymerization, 19
Raman spectra, 74, 77, 83, 85, 86, 87
Raman spectroscopy, 71, 74, 76, 82, 86, 88, 89, 90, 91
reactions, 18
receptors, 11, 13, 15, 17
red wine, 13, 65, 66, 67, 68
redistribution, 13
rejection, 20
researchers, 13
residues, 16, 18, 20, 72, 73
resistance, 51, 66, 73
resolution, 16, 19, 77
response, 11, 51, 52, 68
responsiveness, 4, 5, 36, 45, 48
resveratrol, 51, 52, 53, 59, 66, 67, 68
retardation, viii, 6, 11, 29, 30, 36, 44, 45, 48
risk, 7
room temperature, 75, 76, 78

S

saccharin, 17
salts, ix, 49
scattering, 16
schizophrenia, 15
school, 7
science, 74
secondary metabolism, 66, 69
selective attention, 9
selectivity, 20
selenium, 44
self-assembly, 16
self-esteem, 8
sensitivity, 19
sequencing, 4, 5
serine, 54
serotonin, 15
serum, 3, 6
shape, 87
side chain, vii, 1, 19, 73, 81
signs, 6
silk, 73
silver, 65
Singapore, 48

skin, vii, 1, 53
social competence, 8
social problems, 7
socioeconomic status, 6
sodium, 75
solution, 19, 20, 21, 75, 76, 78, 86
South Africa, 69
Spain, 27, 28, 36, 48, 49, 65
species, 13, 17, 18
spectroscopy, 10, 17, 74, 89
spindle, 72
stability, 47, 72, 73
stabilization, 6
standard deviation, 32, 37, 54, 56, 58, 61, 77
state(s), 3, 16, 46
steel, 19
stimulation, 53
stretching, 83
striatum, 14
structural changes, 82
structure, 10, 17, 72, 73, 74, 82, 85
substitutes, 30, 31, 34, 36, 44, 45
substitution, 4, 30
substrate(s), 59, 69
sulfur, 64, 65
sulfur dioxide, 65
supplementation, 30, 44, 67
surrogates, 18
suspensions, 53
sweeteners, 16, 17, 21
Switzerland, 36
symptoms, 3, 8
synaptic transmission, 11
synaptic vesicles, 13
synthesis, vii, 1, 13, 16, 21, 52, 53, 54, 57, 59, 60, 61, 62
synthetic analogues, 13
syphilis, 3

T

tannins, 53, 66
target population, 4
techniques, 16, 17
temperature, 16, 19
tensile strength, 73, 88
terpenes, 53
therapy, viii, 6, 15, 16, 17, 18, 21, 29, 30, 44, 45, 47
threonine, 51, 54, 61
thymus, 17
tissue, 11, 15
toxicity, 13, 16
trace elements, 44
transformation, 16, 66
transmission, 11
transport, 8, 13
treatment, viii, x, 2, 3, 4, 5, 6, 10, 21, 28, 29, 30, 31, 33, 36, 37, 38, 39, 40, 41, 42, 43, 44, 45, 46, 48, 52, 53, 54, 57, 59, 67, 71, 75, 76, 77, 78, 80, 81, 87, 88
tryptophan, vii, 1, 17, 74, 75, 81, 82, 86, 88
tyramine, 12, 15, 21
tyrosine, vii, viii, 1, 5, 8, 11, 16, 21, 28, 30, 72, 74, 78, 79, 82, 86

U

urea, 53, 63, 65, 66, 67
USA, 23
UV, 51

V

Valencia, 28
valine, 75
variables, ix, 15, 29, 43, 44
varieties, 61, 64, 65, 67
vascular system, 13
vasoconstriction, 13
vegetarian diets, viii, 28, 30
vertebrates, 12
vessels, 13
vibration, 83
vineyard, ix, 50, 51, 52, 53, 57, 62
vinylidene fluoride, 19
viscosity, 17
vitamins, 50
viticulture, ix, 50

W

water, 19, 73, 78
wave number, 82
western blot, 11
white matter, 5
World Health Organization (WHO), 34, 48

Y

yeast, ix, 49, 50, 54, 59, 62, 63, 64, 67

Z

zinc, 44